Advanced Higher

Physics

2004 Exam

2005 Exam

2006 Exam

2007 Exam

2008 Exam

Leckie × Leckie

First exam published in 2004.
Published by Leckie & Leckie Ltd, 3rd Floor, 4 Queen Street, Edinburgh EH2 1JE
tel: 0131 220 6831 fax: 0131 225 9987 enquiries@leckieandleckie.co.uk www.leckieandleckie.co.uk

ISBN 978-1-84372-694-4

A CIP Catalogue record for this book is available from the British Library.

Leckie & Leckie is a division of Huveaux plc.

Leckie & Leckie is grateful to the copyright holders, as credited at the back of the book, for permission to use their material. Every effort has been made to trace the copyright holders and to obtain their permission for the use of copyright material. Leckie & Leckie will gladly receive information enabling them to rectify any error or omission in subsequent editions.

[BLANK PAGE]

X069/701

NATIONAL QUALIFICATIONS 2004

FRIDAY, 28 MAY
1.00 PM – 3.30 PM

PHYSICS
ADVANCED HIGHER

Answer **all** questions.

Any necessary data may be found in the Data Sheet on page two.

Care should be taken to give an appropriate number of significant figures in the final answers to calculations.

Square-ruled paper (if used) should be placed inside the front cover of the answer book for return to the Scottish Qualifications Authority.

DATA SHEET

COMMON PHYSICAL QUANTITIES

Quantity	*Symbol*	*Value*	*Quantity*	*Symbol*	*Value*
Gravitational acceleration on Earth	g	$9{\cdot}8\ \text{m s}^{-2}$	Mass of electron	m_e	$9{\cdot}11 \times 10^{-31}$ kg
Radius of Earth	R_E	$6{\cdot}4 \times 10^{6}$ m	Charge on electron	e	$-1{\cdot}60 \times 10^{-19}$ C
Mass of Earth	M_E	$6{\cdot}0 \times 10^{24}$ kg	Mass of neutron	m_n	$1{\cdot}675 \times 10^{-27}$ kg
Mass of Moon	M_M	$7{\cdot}3 \times 10^{22}$ kg	Mass of proton	m_p	$1{\cdot}673 \times 10^{-27}$ kg
Mean Radius of Moon Orbit		$3{\cdot}84 \times 10^{8}$ m	Planck's constant	h	$6{\cdot}63 \times 10^{-34}$ J s
Universal constant of gravitation	G	$6{\cdot}67 \times 10^{-11}\ \text{m}^3\ \text{kg}^{-1}\ \text{s}^{-2}$	Permittivity of free space	ε_0	$8{\cdot}85 \times 10^{-12}\ \text{F m}^{-1}$
Speed of light in vacuum	c	$3{\cdot}0 \times 10^{8}\ \text{m s}^{-1}$	Permeability of free space	μ_0	$4\pi \times 10^{-7}\ \text{H m}^{-1}$
Speed of sound in air	v	$3{\cdot}4 \times 10^{2}\ \text{m s}^{-1}$			

REFRACTIVE INDICES

The refractive indices refer to sodium light of wavelength 589 nm and to substances at a temperature of 273 K.

Substance	*Refractive index*	*Substance*	*Refractive index*
Diamond	2·42	Glycerol	1·47
Glass	1·51	Water	1·33
Ice	1·31	Air	1·00
Perspex	1·49	Magnesium Fluoride	1·38

SPECTRAL LINES

Element	*Wavelength*/nm	*Colour*
Hydrogen	656	Red
	486	Blue-green
	434	Blue-violet
	410	Violet
	397	Ultraviolet
	389	Ultraviolet
Sodium	589	Yellow

Element	*Wavelength*/nm	*Colour*
Cadmium	644	Red
	509	Green
	480	Blue

Lasers

Element	*Wavelength*/nm	*Colour*
Carbon dioxide	9550 10590	Infrared
Helium-neon	633	Red

PROPERTIES OF SELECTED MATERIALS

Substance	*Density*/ kg m^{-3}	*Melting Point*/ K	*Boiling Point*/ K	*Specific Heat Capacity*/ J kg^{-1} K^{-1}	*Specific Latent Heat of Fusion*/ J kg^{-1}	*Specific Latent Heat of Vaporisation*/ J kg^{-1}
Aluminium	$2{\cdot}70 \times 10^3$	933	2623	$9{\cdot}02 \times 10^2$	$3{\cdot}95 \times 10^5$	....
Copper	$8{\cdot}96 \times 10^3$	1357	2853	$3{\cdot}86 \times 10^2$	$2{\cdot}05 \times 10^5$	....
Glass	$2{\cdot}60 \times 10^3$	1400	....	$6{\cdot}70 \times 10^2$	....	....
Ice	$9{\cdot}20 \times 10^2$	273	....	$2{\cdot}10 \times 10^3$	$3{\cdot}34 \times 10^5$	....
Glycerol	$1{\cdot}26 \times 10^3$	291	563	$2{\cdot}43 \times 10^3$	$1{\cdot}81 \times 10^5$	$8{\cdot}30 \times 10^5$
Methanol	$7{\cdot}91 \times 10^2$	175	338	$2{\cdot}52 \times 10^3$	$9{\cdot}9 \times 10^4$	$1{\cdot}12 \times 10^6$
Sea Water	$1{\cdot}02 \times 10^3$	264	377	$3{\cdot}93 \times 10^3$	....	....
Water	$1{\cdot}00 \times 10^3$	273	373	$4{\cdot}19 \times 10^3$	$3{\cdot}34 \times 10^5$	$2{\cdot}26 \times 10^6$
Air	1·29	....	....	....	....	....
Hydrogen	$9{\cdot}0 \times 10^{-2}$	14	20	$1{\cdot}43 \times 10^4$	....	$4{\cdot}50 \times 10^5$
Nitrogen	1·25	63	77	$1{\cdot}04 \times 10^3$	....	$2{\cdot}00 \times 10^5$
Oxygen	1·43	55	90	$9{\cdot}18 \times 10^2$	....	$2{\cdot}40 \times 10^5$

The gas densities refer to a temperature of 273 K and a pressure of $1{\cdot}01 \times 10^5$ Pa.

Marks

1. The relativistic mass m of a moving object is given by

$$m = \frac{m_0}{\sqrt{(1 - \frac{v^2}{c^2})}}$$

where the symbols have their usual meanings.

(*a*) Calculate the speed at which the relativistic mass of an object is equal to three times its rest mass. **2**

(*b*) An electron is emitted with a speed of $0 \cdot 90\,c$ from a radioactive nucleus.

Calculate the relativistic energy of this electron. **3**

(5)

[Turn over

Marks

2. A yo-yo consists of two discs mounted on an axle.

A length of string is attached to the axle and wound round the axle.

With the string fully wound, the yo-yo is suspended from a horizontal support as shown in Figure 1(*a*).

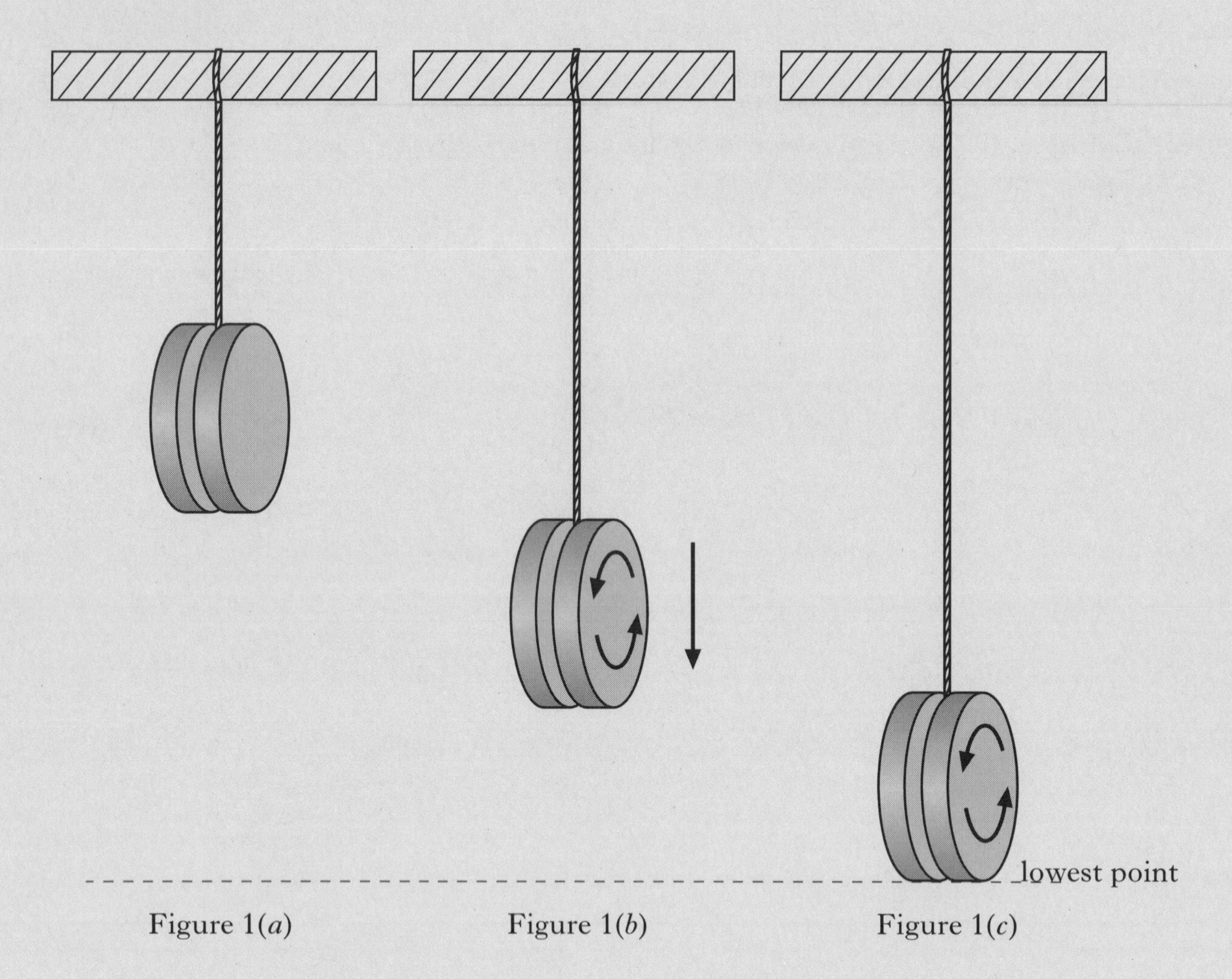

Figure 1(*a*) Figure 1(*b*) Figure 1(*c*)

The yo-yo is released from rest and rotates as it falls, as shown in Figure 1(*b*). The string is fully unwound at the yo-yo's lowest point, as shown in Figure 1(*c*). The yo-yo then rises, rewinding the string.

(*a*) State the type(s) of energy which the yo-yo has when it is at the position shown in:

(i) Figure 1(*b*);

(ii) Figure 1(*c*). **2**

(*b*) Each disc has a mass m of 0·100 kg and a radius r of 0·050 m.

The moment of inertia of a disc is given by $\frac{1}{2}mr^2$.

The moment of inertia of the axle is negligible.

Calculate the moment of inertia of the yo-yo. **2**

(*c*) When the yo-yo is at the position shown in Figure 1(*c*) it has an angular velocity of 120 rad s^{-1}.

Calculate the maximum height to which the yo-yo could rise as it rewinds the string. **2**

Marks

2. (continued)

(*d*) One type of yo-yo has four friction pads inside each disc. Each friction pad is held in place by a spring which exerts a force of 5·00 N. At low angular velocities the friction pads grip the axle as shown in Figure 2.

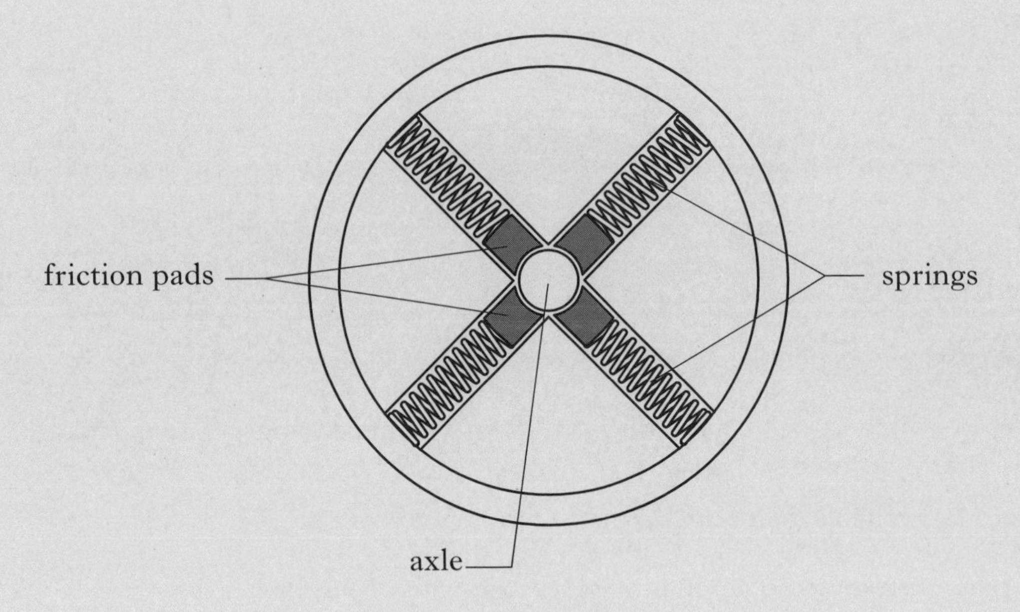

Figure 2

At higher angular velocities the pads move away from the axle and compress the springs. This releases the axle and allows the discs to spin freely.

(i) Explain why the friction pads move away from the axle.

(ii) Each friction pad can be considered as a point mass of 12·0 g at a radius of 10·0 mm from the centre of the axle.

Calculate the minimum angular velocity at which the axle is released from the friction pads. **3**

(9)

[Turn over

Marks

3. A grinder is used for cutting paving slabs.

The grinder has a motor and a disc with an abrasive edge as shown in Figure 3.

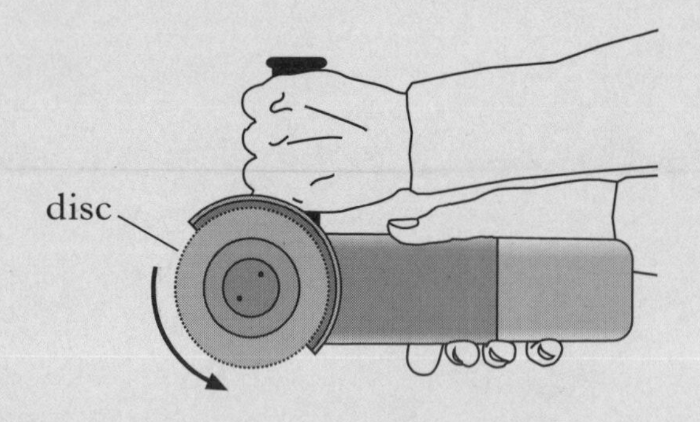

Figure 3

The motor is switched on and the disc reaches a maximum angular velocity of 600 revolutions per minute.

The motor is switched off and the disc slows uniformly to rest in 30 s.

(*a*) Calculate the maximum angular velocity of the disc in rad s^{-1}. **1**

(*b*) Calculate the angular acceleration of the disc as it slows. **2**

(*c*) How many revolutions does the disc make during this time? **3**

(*d*) The moment of inertia of the disc is $2{\cdot}16 \times 10^{-3}$ kg m^2.
Calculate the torque acting on the disc as it slows. **2**

(*e*) The disc is replaced by a disc of **half** the radius and **double** the mass. The motor is switched on and this disc also reaches a maximum angular velocity of 600 revolutions per minute. The grinder is switched off and a torque equal to that in part (*d*) acts on the disc.

Explain whether this disc comes to rest in a time greater than, equal to or less than 30 s.

The moment of inertia of a disc is given by $\frac{1}{2}mr^2$. **2**

(10)

Marks

4. The gravitational pull of the Earth keeps a satellite in a circular orbit.

(*a*) Show that for an orbit of radius r the period T is given by

$$T = 2\pi\sqrt{\frac{r^3}{GM_E}}$$

where the symbols have their usual meanings. **2**

(*b*) A polar orbiting satellite is used to map the Earth by photographing strips of the surface as it orbits, as shown in Figure 4.

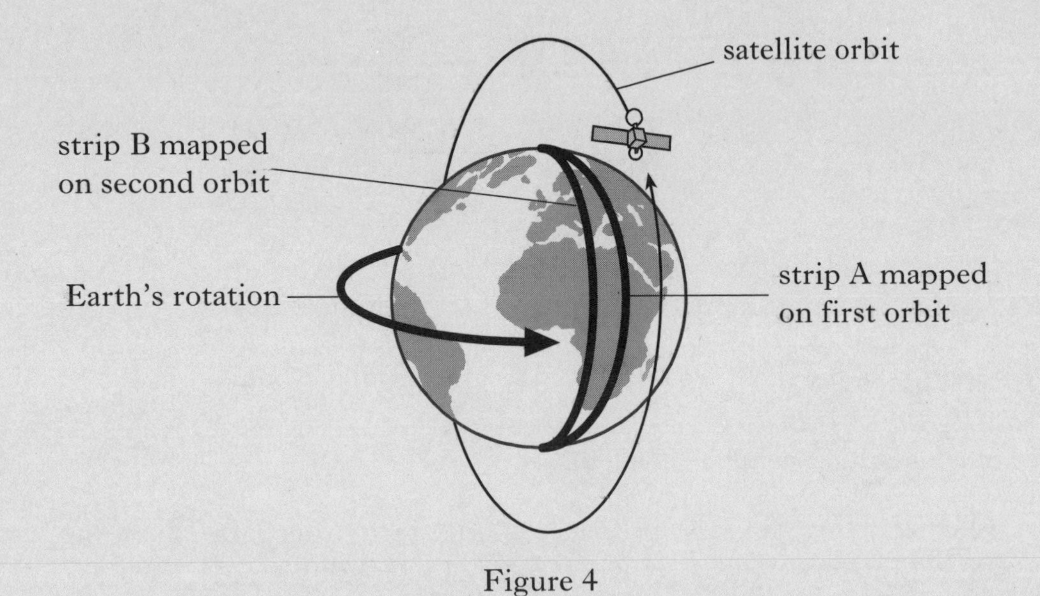

Figure 4

The plane of the satellite orbit is fixed. The Earth rotates and so the satellite maps a different strip on each orbit.

(i) The satellite orbits at a height of 80 km above the surface of the Earth. Assuming the Earth to be spherical, show that the period of the orbit is approximately 86 minutes.

(ii) The Earth's angular velocity is $7{\cdot}3 \times 10^{-5}$ rad s^{-1}.

Calculate the distance along the equator between strips A and B which are mapped on consecutive orbits. **4**

(6)

[Turn over

Marks

5. A mass of 0·50 kg is suspended from a spring of negligible mass, as shown in Figure 5(a).

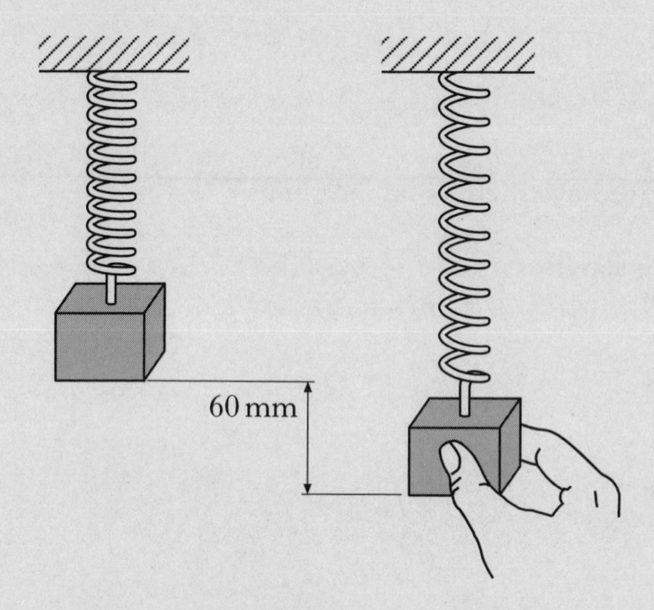

Figure 5(a) Figure 5(b)

A student pulls the mass down a distance of 60 mm and holds it in the position shown in Figure 5(b).

The tension in the spring is now 7·0 N.

(a) By considering the vertical forces acting on the mass, calculate the force applied by the student to hold the mass in this position. **1**

(b) The student releases the mass, which performs simple harmonic motion.

(i) State the relationship between the unbalanced force acting on the mass and the displacement of the mass.

(ii) Calculate the acceleration of the mass immediately after its release.

(iii) State the initial amplitude of the oscillations. **4**

(c) The oscillations of the mass are described by the equation

$$\frac{d^2y}{dt^2} = -\omega^2 y.$$

(i) Name the physical quantity represented by the term $\frac{d^2y}{dt^2}$.

(ii) Calculate the frequency of the oscillations. **3**

(8)

Marks

6. (*a*) (i) Define the term *electric field strength*.

(ii) Two parallel plates are separated by distance d. The potential difference between the plates is V.

Derive the expression for the electric field strength E between the plates in terms of V and d. **3**

(*b*) The electric field pattern between two parallel metal plates is shown in Figure 6.

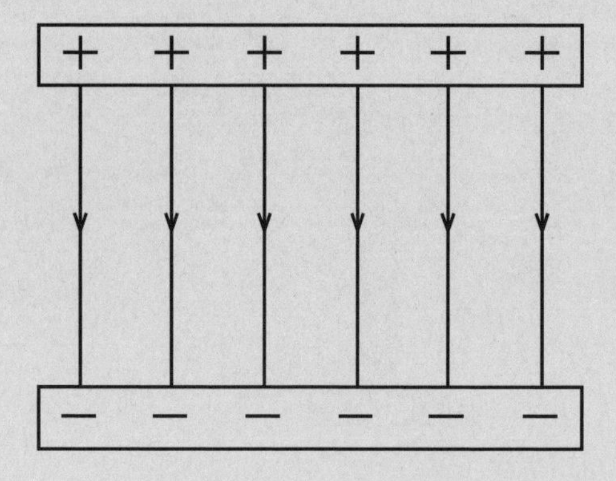

Figure 6

An uncharged, conducting sphere is placed between the plates as shown in Figure 7.

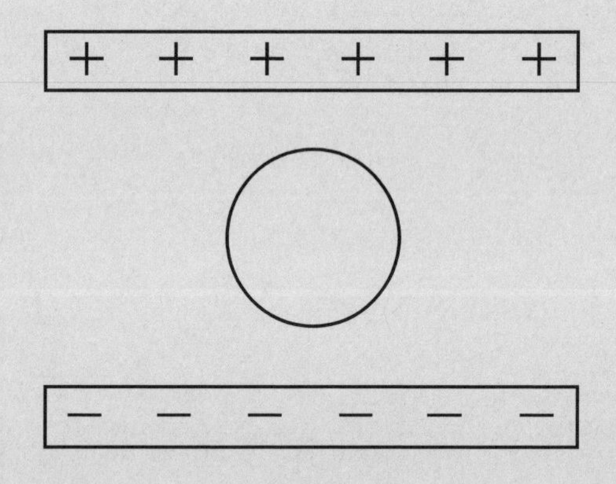

Figure 7

(i) Copy and complete Figure 7 to show the electric field pattern between the plates.

(ii) On your diagram, show the charge distribution on the sphere.

(iii) State the value of the electric field strength inside the sphere. **4**

(7)

[Turn over

Marks

7. (*a*) State Coulomb's law for the electrostatic force between two point charges. **1**

(*b*) The two identical conducting spheres R and S, shown in Figure 8, are initially uncharged.

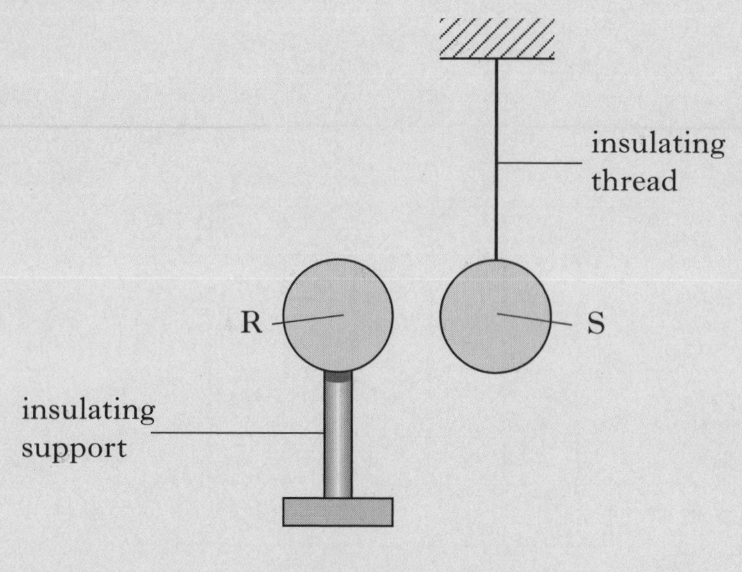

Figure 8

Describe how sphere R can be given a positive charge and sphere S an equal negative charge by induction, using a positively charged insulating rod. **2**

(*c*) Two identical conducting spheres X and Y shown in Figure 9 have equal and opposite charges.

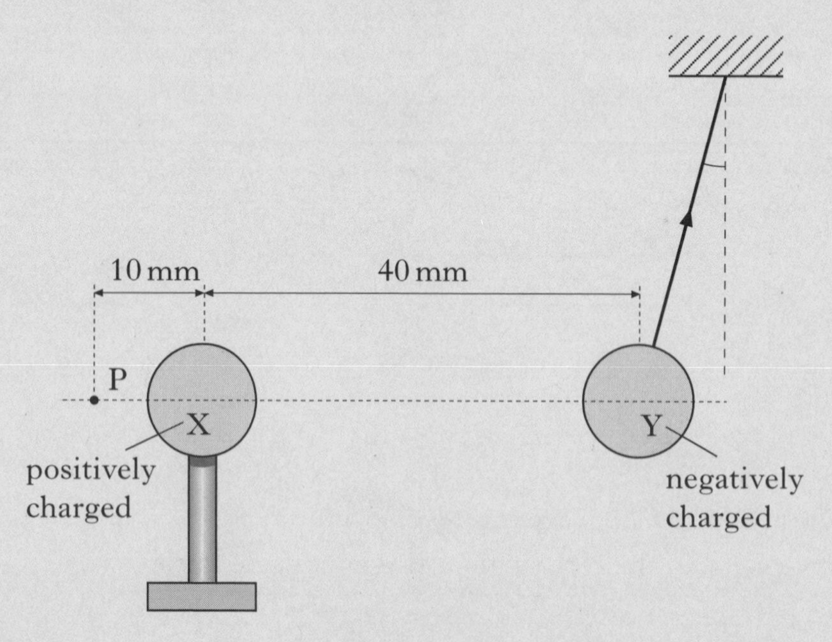

Figure 9

(i) The force between the spheres is $3{\cdot}0 \times 10^{-5}$ N.

By considering the spheres as point charges separated by a distance of 40 mm, show that the charge on each sphere is $2{\cdot}3$ nC.

(ii) Calculate the electrostatic potential at point P, 10 mm from X, as shown in Figure 9.

Marks

7. (*c*) (continued)

(iii) In reality, the spheres are not point charges.

Draw a sketch to show how charge is distributed on each sphere when the spheres are in the positions shown in Figure 9. **6**

(*d*) Sphere Y has mass $2{\cdot}5 \times 10^{-5}$ kg and hangs at an angle α to the vertical, as shown in Figure 10. The horizontal force acting on the sphere is $3{\cdot}0 \times 10^{-5}$ N.

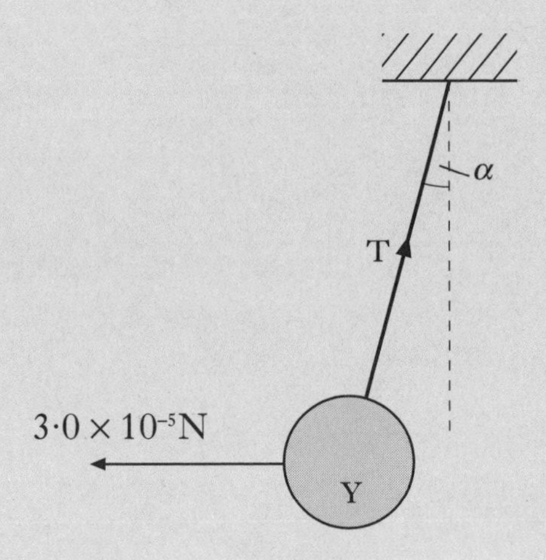

Figure 10

The mass of the thread is negligible.

(i) Calculate the tension T in the thread.

(ii) Calculate the size of angle α. **4**

(13)

[Turn over

Marks

8. A particle of mass m and charge q is fired with speed v into a magnetic field of uniform magnetic induction B. The particle enters the field at point X and follows a semicircular path, before leaving the field at point Y, as shown in Figure 11.

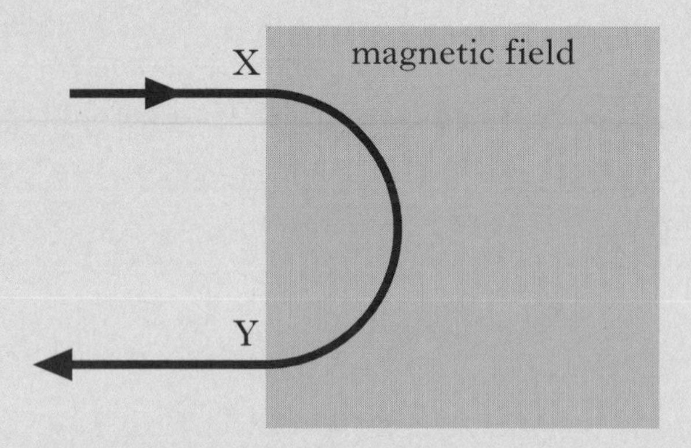

Figure 11

(*a*) Show that the radius r of the semicircular path is given by

$$r = \frac{mv}{qB}.$$ **2**

(*b*) Using the above relationship, show that the time taken for the particle to follow the semicircular path in the magnetic field is independent of the speed of the particle. **2**

(*c*) An electron with speed $2{\cdot}0 \times 10^6\,\mathrm{m\,s^{-1}}$ is fired, as shown in Figure 11, into a magnetic field of uniform magnetic induction $5{\cdot}0\,\mathrm{mT}$. Calculate the time during which the electron is in the magnetic field. **2**

(6)

Marks

9. An inductor of negligible resistance is connected in the circuit shown in Figure 12.

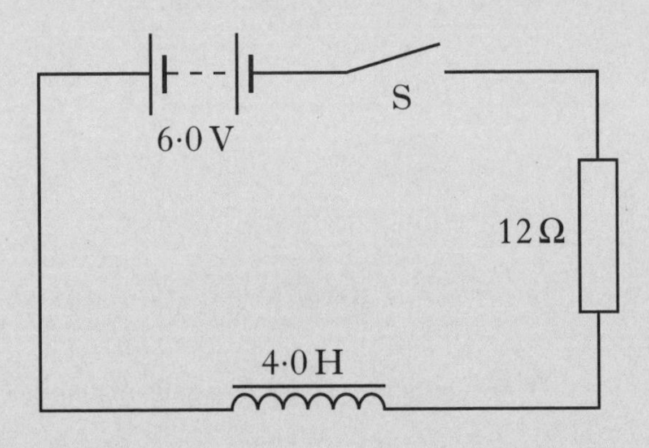

Figure 12

(*a*) Sketch a graph to show how the potential difference across the resistor varies with time after switch S is closed. A numerical scale is required on the potential difference axis. **2**

(*b*) At a point in time after switch S is closed, the current in the circuit is 0·20 A. Calculate the rate of change of current at this time. **3**

(5)

[Turn over

Marks

10. Two long parallel conductors, distance r apart, carry currents I_1 and I_2.

(*a*) Show that the force per unit length acting on each conductor is given by

$$\frac{F}{l} = \frac{\mu_0 I_1 I_2}{2\pi r}$$

where the symbols have their usual meanings. **2**

(*b*) In some countries direct current is used for transmitting power over long distances. Two direct current transmission cables each carry a current of 850 A. The currents are in opposite directions as shown in Figure 13.

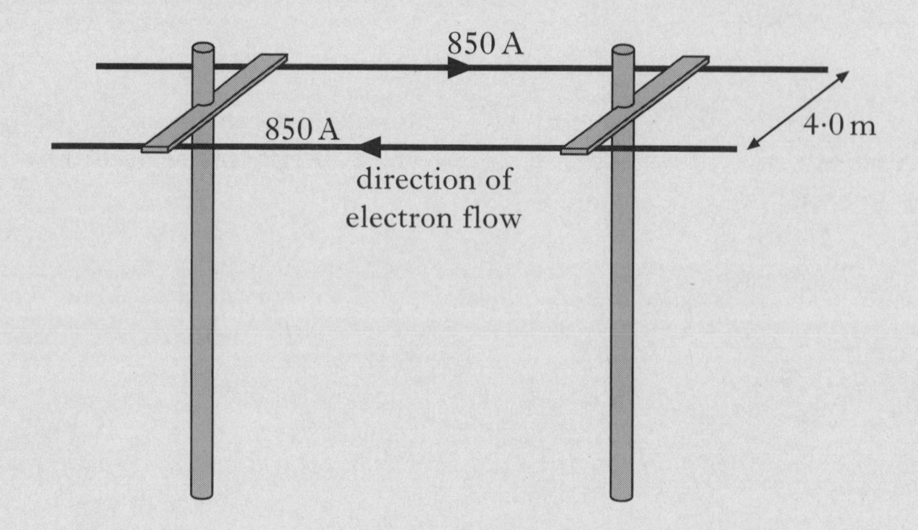

Figure 13

The cables are parallel and are separated by a distance of 4·0 m.

(i) Calculate the force per unit length between the cables due to the currents in the cables.

(ii) Does this force tend to move the cables together or apart?

(iii) Determine the magnitude **and** direction of the resultant magnetic induction, due to both cables, at a point midway between the cables. **5**

Marks

10. (continued)

(*c*) The direction of the Earth's magnetic field is at an angle of 60 degrees to one cable, as shown in Figure 14.

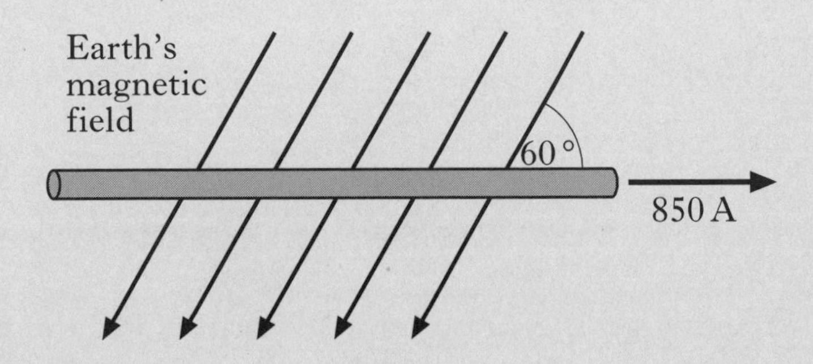

Figure 14

In the region of the cable, the magnetic induction of the Earth's field is 52 μT. Calculate the force per unit length on this cable due to the Earth's magnetic field and the current in this cable. **2**

(9)

[Turn over

Marks

11. A transverse wave is described by the expression

$$y = 8{\cdot}0 \sin(12t - 0{\cdot}50x)$$

where t is in seconds and x and y are in metres.

(*a*) For this wave, calculate the:

(i) frequency;

(ii) wavelength. **2**

(*b*) (i) Calculate the phase difference, in radians, between the point at $x = 3{\cdot}0$ m and the point at $x = 4{\cdot}0$ m.

(ii) Calculate the time for the wave to travel between these two points. **4**

(*c*) The wave is reflected and loses some energy.

State a possible equation for the reflected wave. **2**

(8)

Marks

12. A television aerial is shown in Figure 15.

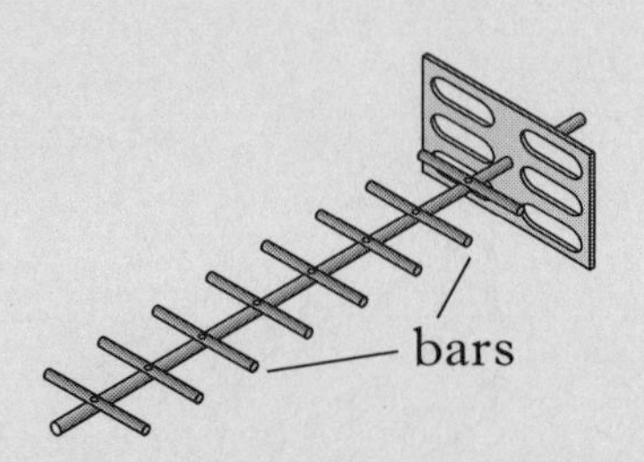

Figure 15

(*a*) Instructions for installing the aerial state

"*The television waves received are plane polarised. The aerial does not pick up a strong signal unless the plane of the bars is the same as the plane of polarisation of the television waves.*"

(i) Explain the term *plane polarised.*

(ii) The aerial is installed and connected to a television.

The television has a clear picture when the bars of the aerial are horizontal as shown in Figure 15.

The aerial is now slowly rotated until the bars are vertical as shown in Figure 16.

Describe what happens to the television picture during this rotation. **2**

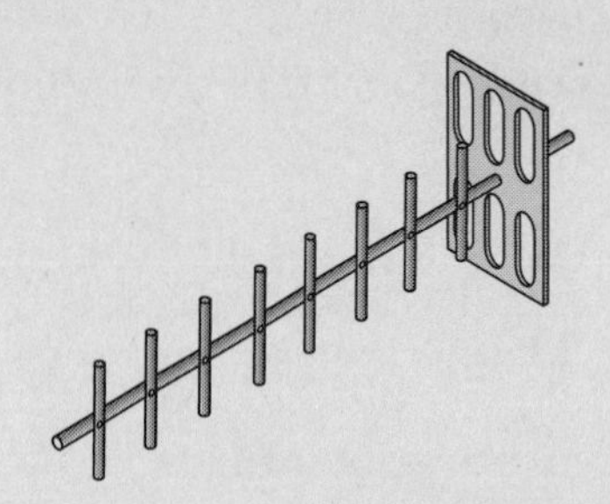

Figure 16

(*b*) Unpolarised light strikes the surface of a transparent material at the Brewster angle i_p, as shown in Figure 17. The reflected light is plane polarised.

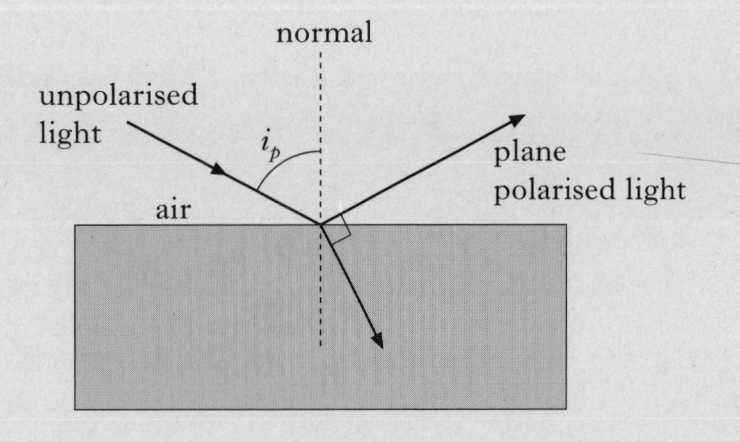

Figure 17

(i) Derive the expression

$$n = \tan i_p$$

where n is the refractive index of the transparent material.

(ii) Calculate the Brewster angle for perspex. **4**

(6)

Marks

13. A student sets up a "Young's double slit" experiment, as shown in Figure 18, to measure the wavelength of laser light.

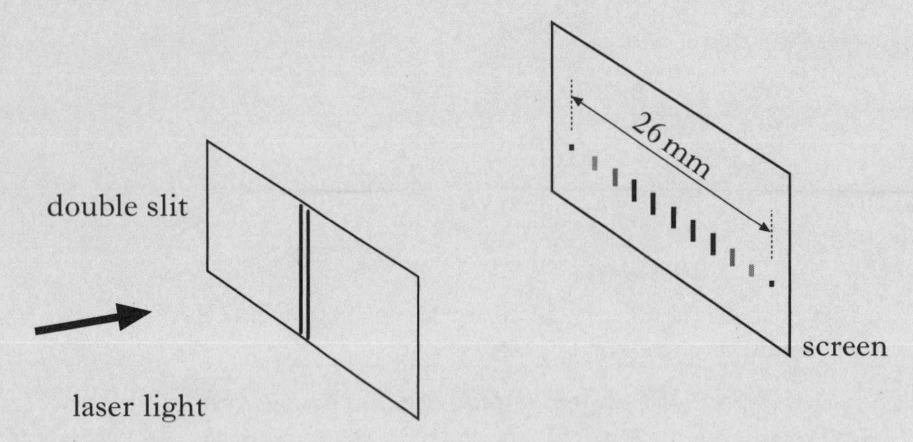

Figure 18

The student obtains the following results.

Separation of 11 fringes	26 (±2) mm
Distance to screen from slits	2·00 (±0·01) m
Separation of slits	0·52 (±0·02) mm

(*a*) Calculate the wavelength of the laser light. **2**

(*b*) Calculate the percentage uncertainty in this wavelength. **3**

(*c*) Suggest an improvement to the experiment that would reduce the uncertainty in the calculated value of the wavelength.
Justify your answer. **2**

(*d*) Which principle does this experiment illustrate, interference by division of wavefront or by division of amplitude? **1**

(8)

[*END OF QUESTION PAPER*]

2005 | Advanced Higher

[BLANK PAGE]

X069/701

NATIONAL
QUALIFICATIONS
2005

TUESDAY, 24 MAY
1.00 PM – 3.30 PM

PHYSICS
ADVANCED HIGHER

Answer **all** questions.

Any necessary data may be found in the Data Sheet on page two.

Care should be taken to give an appropriate number of significant figures in the final answers to calculations.

Square-ruled paper (if used) should be placed inside the front cover of the answer book for return to the Scottish Qualifications Authority.

LIB X069/701 6/3/4970

DATA SHEET

COMMON PHYSICAL QUANTITIES

Quantity	*Symbol*	*Value*	*Quantity*	*Symbol*	*Value*
Gravitational acceleration on Earth	g	$9{\cdot}8\ \text{m s}^{-2}$	Mass of electron	m_e	$9{\cdot}11 \times 10^{-31}$ kg
Radius of Earth	R_E	$6{\cdot}4 \times 10^{6}$ m	Charge on electron	e	$-1{\cdot}60 \times 10^{-19}$ C
Mass of Earth	M_E	$6{\cdot}0 \times 10^{24}$ kg	Mass of neutron	m_n	$1{\cdot}675 \times 10^{-27}$ kg
Mass of Moon	M_M	$7{\cdot}3 \times 10^{22}$ kg	Mass of proton	m_p	$1{\cdot}673 \times 10^{-27}$ kg
Mean Radius of Moon Orbit		$3{\cdot}84 \times 10^{8}$m	Mass of alpha particle	m_α	$6{\cdot}645 \times 10^{-27}$ kg
Universal constant of gravitation	G	$6{\cdot}67 \times 10^{-11}\ \text{m}^3\ \text{kg}^{-1}\ \text{s}^{-2}$	Charge on alpha particle		$3{\cdot}20 \times 10^{-19}$ C
Speed of light in vacuum	c	$3{\cdot}0 \times 10^{8}\ \text{m s}^{-1}$	Planck's constant	h	$6{\cdot}63 \times 10^{-34}$ J s
Speed of sound in air	v	$3{\cdot}4 \times 10^{2}\ \text{m s}^{-1}$	Permittivity of free space	ε_0	$8{\cdot}85 \times 10^{-12}\ \text{F m}^{-1}$
			Permeability of free space	μ_0	$4\pi \times 10^{-7}\ \text{H m}^{-1}$

REFRACTIVE INDICES

The refractive indices refer to sodium light of wavelength 589 nm and to substances at a temperature of 273 K.

Substance	*Refractive index*	*Substance*	*Refractive index*
Diamond	2·42	Glycerol	1·47
Glass	1·51	Water	1·33
Ice	1·31	Air	1·00
Perspex	1·49	Magnesium Fluoride	1·38

SPECTRAL LINES

Element	*Wavelength*/nm	*Colour*
Hydrogen	656	Red
	486	Blue-green
	434	Blue-violet
	410	Violet
	397	Ultraviolet
	389	Ultraviolet
Sodium	589	Yellow

Element	*Wavelength*/nm	*Colour*
Cadmium	644	Red
	509	Green
	480	Blue

Lasers

Element	*Wavelength*/nm	*Colour*
Carbon dioxide	9550, 10590	Infrared
Helium-neon	633	Red

PROPERTIES OF SELECTED MATERIALS

Substance	*Density*/ kg m^{-3}	*Melting Point*/ K	*Boiling Point*/ K	*Specific Heat Capacity*/ J kg^{-1} K^{-1}	*Specific Latent Heat of Fusion*/ J kg^{-1}	*Specific Latent Heat of Vaporisation*/ J kg^{-1}
Aluminium	$2{\cdot}70 \times 10^3$	933	2623	$9{\cdot}02 \times 10^2$	$3{\cdot}95 \times 10^5$	....
Copper	$8{\cdot}96 \times 10^3$	1357	2853	$3{\cdot}86 \times 10^2$	$2{\cdot}05 \times 10^5$	....
Glass	$2{\cdot}60 \times 10^3$	1400	....	$6{\cdot}70 \times 10^2$	....	....
Ice	$9{\cdot}20 \times 10^2$	273	....	$2{\cdot}10 \times 10^3$	$3{\cdot}34 \times 10^5$	....
Glycerol	$1{\cdot}26 \times 10^3$	291	563	$2{\cdot}43 \times 10^3$	$1{\cdot}81 \times 10^5$	$8{\cdot}30 \times 10^5$
Methanol	$7{\cdot}91 \times 10^2$	175	338	$2{\cdot}52 \times 10^3$	$9{\cdot}9 \times 10^4$	$1{\cdot}12 \times 10^6$
Sea Water	$1{\cdot}02 \times 10^3$	264	377	$3{\cdot}93 \times 10^3$	....	....
Water	$1{\cdot}00 \times 10^3$	273	373	$4{\cdot}19 \times 10^3$	$3{\cdot}34 \times 10^5$	$2{\cdot}26 \times 10^6$
Air	1·29	....	....	....	....	....
Hydrogen	$9{\cdot}0 \times 10^{-2}$	14	20	$1{\cdot}43 \times 10^4$	....	$4{\cdot}50 \times 10^5$
Nitrogen	1·25	63	77	$1{\cdot}04 \times 10^3$	....	$2{\cdot}00 \times 10^5$
Oxygen	1·43	55	90	$9{\cdot}18 \times 10^2$	....	$2{\cdot}40 \times 10^5$

The gas densities refer to a temperature of 273 K and a pressure of $1{\cdot}01 \times 10^5$ Pa.

Marks

1. A compact disc (CD) stores information on the surface as shown in Figure 1.

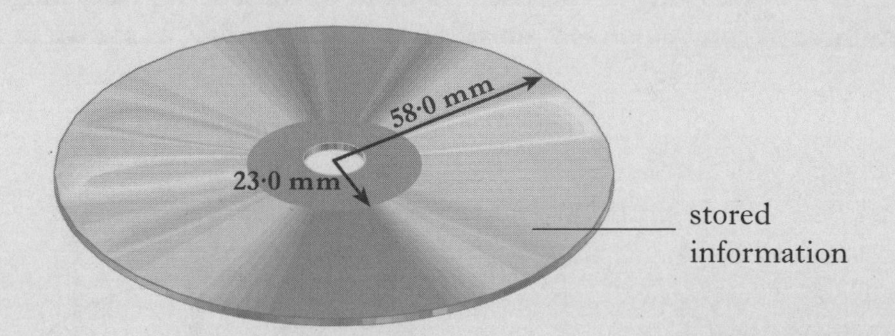

Figure 1

The information is retrieved by an optical reader which moves outwards as the CD rotates, as shown in Figure 2.

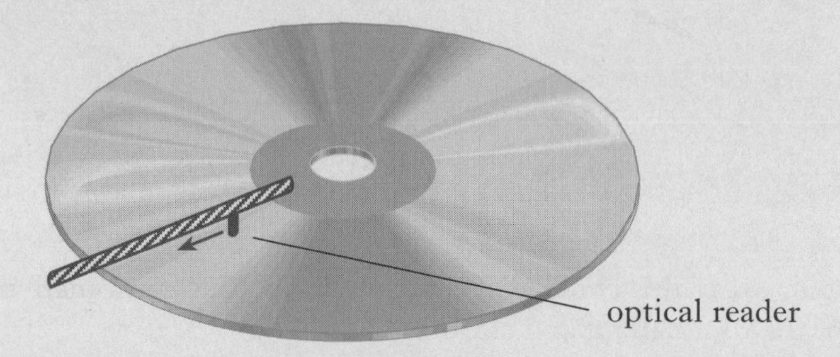

Figure 2

The part of the CD below the reader must always have a tangential speed of $1 \cdot 30\,\mathrm{m\,s^{-1}}$.

(*a*) The reader starts at a radius of 23·0 mm from the centre of the CD. Calculate the angular velocity of the CD at the start. **2**

(*b*) Show that the CD rotates at $22 \cdot 4\,\mathrm{rad\,s^{-1}}$ when the reader reaches the outer edge of the disc. **1**

(*c*) Explain why the angular velocity of the CD decreases as the CD plays. **1**

(*d*) The CD makes a total of $2 \cdot 80 \times 10^4$ revolutions from start to finish.

(i) Show that the total angular displacement of the CD is $1 \cdot 76 \times 10^5$ radians. **1**

(ii) Calculate the average angular acceleration of the CD as the disc is played from start to finish. **2**

(iii) Calculate the total playing time of the CD. **2**

(9)

[Turn over

Marks

2. A playground roundabout has a radius of 2·0 m and a moment of inertia of 500 kg m^2 about its axis of rotation. A child of mass 25 kg runs tangentially to the stationary roundabout and jumps on to its outer edge, as shown in Figure 3.

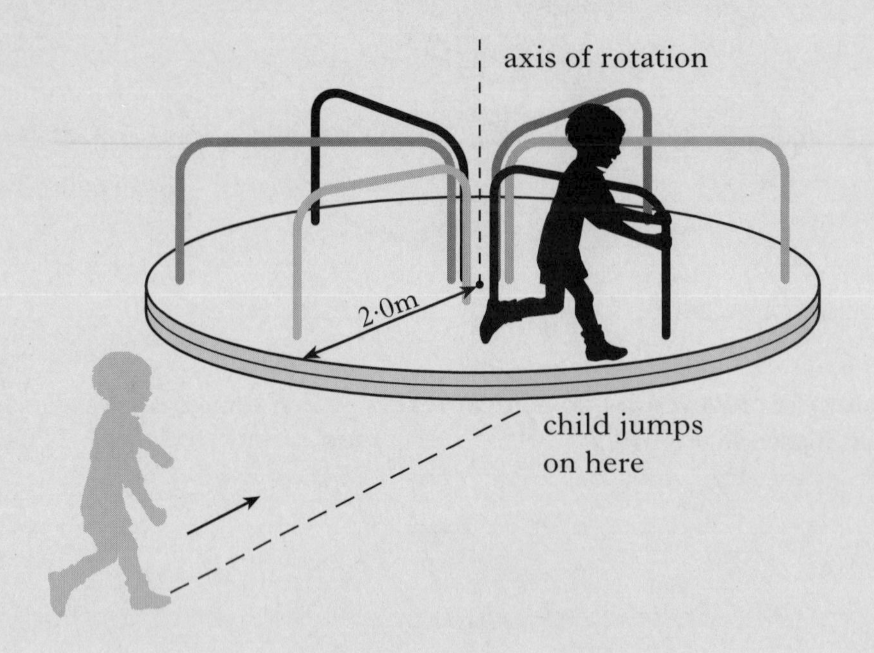

Figure 3

(*a*) Show that, with the child at the outer edge, the combined moment of inertia of the roundabout and child is 600 kg m^2. **2**

(*b*) State what is meant by *conservation of angular momentum*. **1**

(*c*) At the point of jumping on to the roundabout, the tangential speed of the child is 2·4 m s^{-1}. At this **point**, calculate:

(i) the linear momentum of the child; **1**

(ii) the angular momentum of the child about the axis of rotation of the roundabout. **1**

(*d*) Calculate the angular velocity of the roundabout and child just after the child jumps on. **2**

(*e*) Calculate the loss of kinetic energy as the child jumps on to the roundabout. **2**

(*f*) The roundabout with the child onboard makes half a complete revolution before coming to rest.

Calculate the frictional torque acting on the roundabout. **3**

(12)

Marks

3. (*a*) (i) A satellite orbits a planet of mass M. The orbital radius of the satellite is R and the orbital period is T.

Show that

$$T^2 = \frac{4\pi^2 R^3}{GM}$$

2

(ii) Calculate the time taken by the Moon to make one complete orbit of the Earth. **2**

(*b*) A satellite orbits 400 km above the Earth's surface as shown in Figure 4.

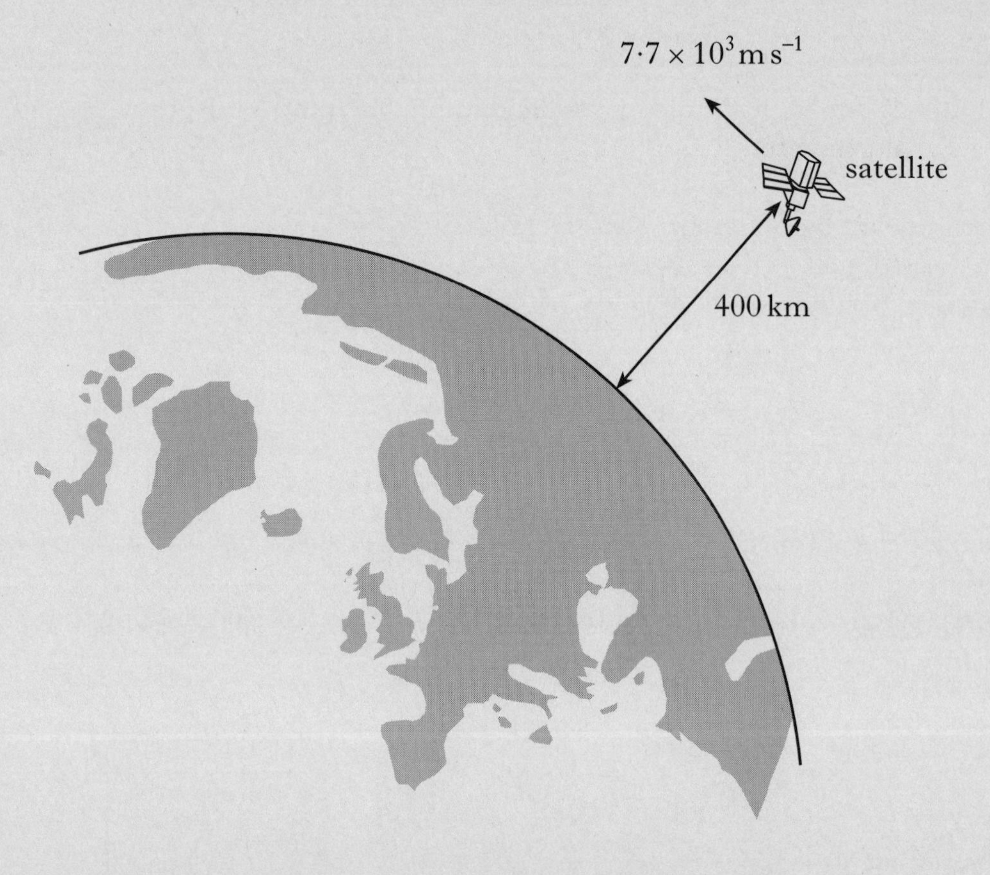

Figure 4

The satellite has a mass of 900 kg and a speed of $7{\cdot}7 \times 10^3\,\text{m s}^{-1}$.

(i) Show that the potential energy of the satellite is $-5{\cdot}3 \times 10^{10}$ J. **2**

(ii) Calculate the total energy of the satellite. **2**

(8)

[Turn over

Marks

4. The flexible paper cone of a loudspeaker vibrates, producing a sound. The loudspeaker has a small cap at the centre of the cone as shown in Figure 5.

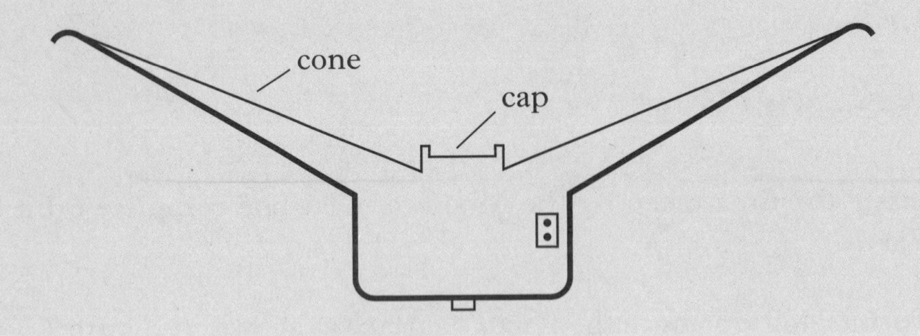

Figure 5

The cone and cap vibrate with simple harmonic motion when the loudspeaker is connected to a signal generator.

(*a*) State what is meant by *simple harmonic motion*. **1**

(*b*) At a particular frequency, the velocity of the cap, in $m\,s^{-1}$, is given by the expression

$$v = 0{\cdot}50 \cos 625t.$$

(i) Calculate the frequency of the sound emitted by the loudspeaker. **2**

(ii) Calculate the amplitude of vibration of the loudspeaker cap. **2**

(*c*) A small polystyrene sphere is placed on the cap of the loudspeaker as shown in Figure 6.

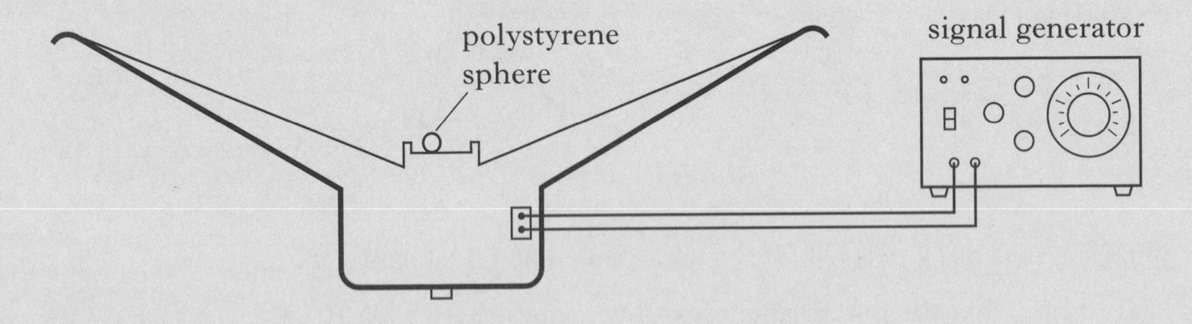

Figure 6

The frequency of the signal generator is slowly increased from zero. At low frequencies the polystyrene sphere stays in contact with the cap. At one particular frequency the sphere just loses contact with the cap. State the maximum acceleration of the cap when this occurs. Justify your answer. **2**

(7)

Marks

5. (*a*) Two point charges with values $+4{\cdot}0\,\mu C$ and $-6{\cdot}0\,\mu C$ are placed $5{\cdot}0\,mm$ apart. Point X lies on the line between the charges as shown in Figure 7.

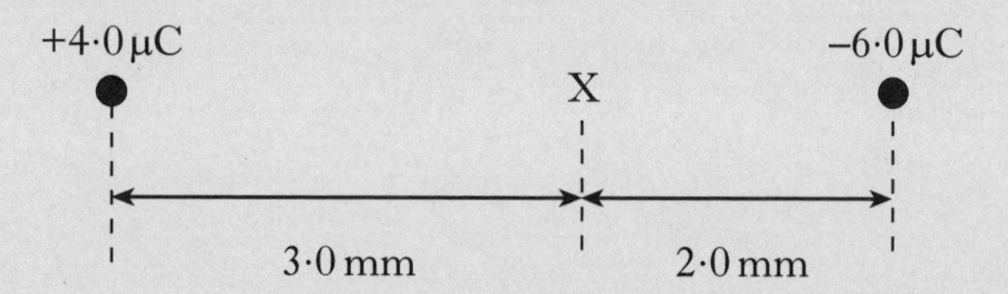

Figure 7

(i) Calculate the magnitude of the electric field strength at point X. **2**

(ii) State the direction of the electric field at point X. **1**

(*b*) A hollow uncharged metal cylinder is placed midway between two parallel plates which are connected to a d.c. power supply as shown in Figure 8.

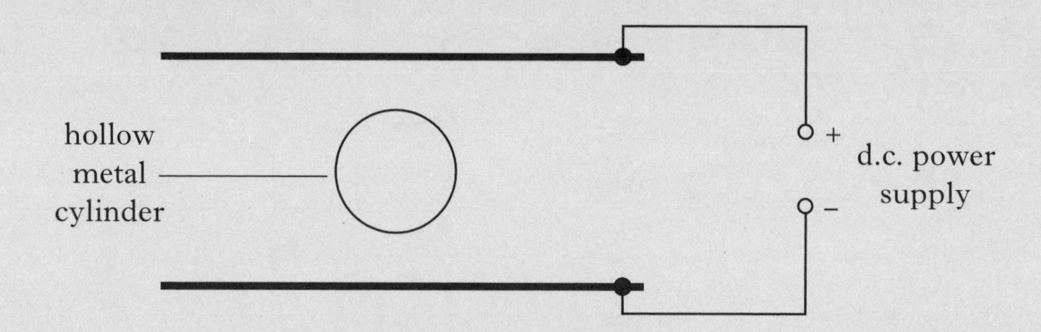

Figure 8

(i) Copy and complete the above diagram showing:

(A) the electric field lines in the space between the parallel plates; **2**

(B) the charge distribution induced on the cylinder. **1**

(ii) Coaxial cable consists of a central wire surrounded by a metal mesh, as shown in Figure 9.

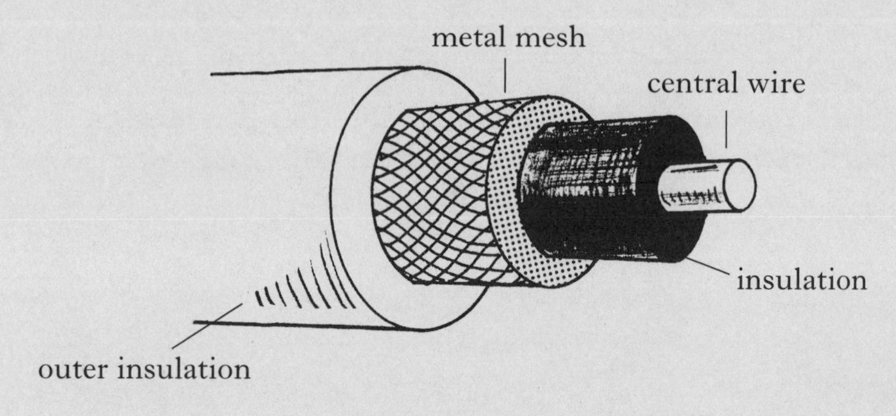

Figure 9

Explain why coaxial cable is designed in this way. **1**

(7)

Marks

6. In an evacuated tube a beam of electrons is deflected by an electric field between two parallel plates as shown in Figure 10.

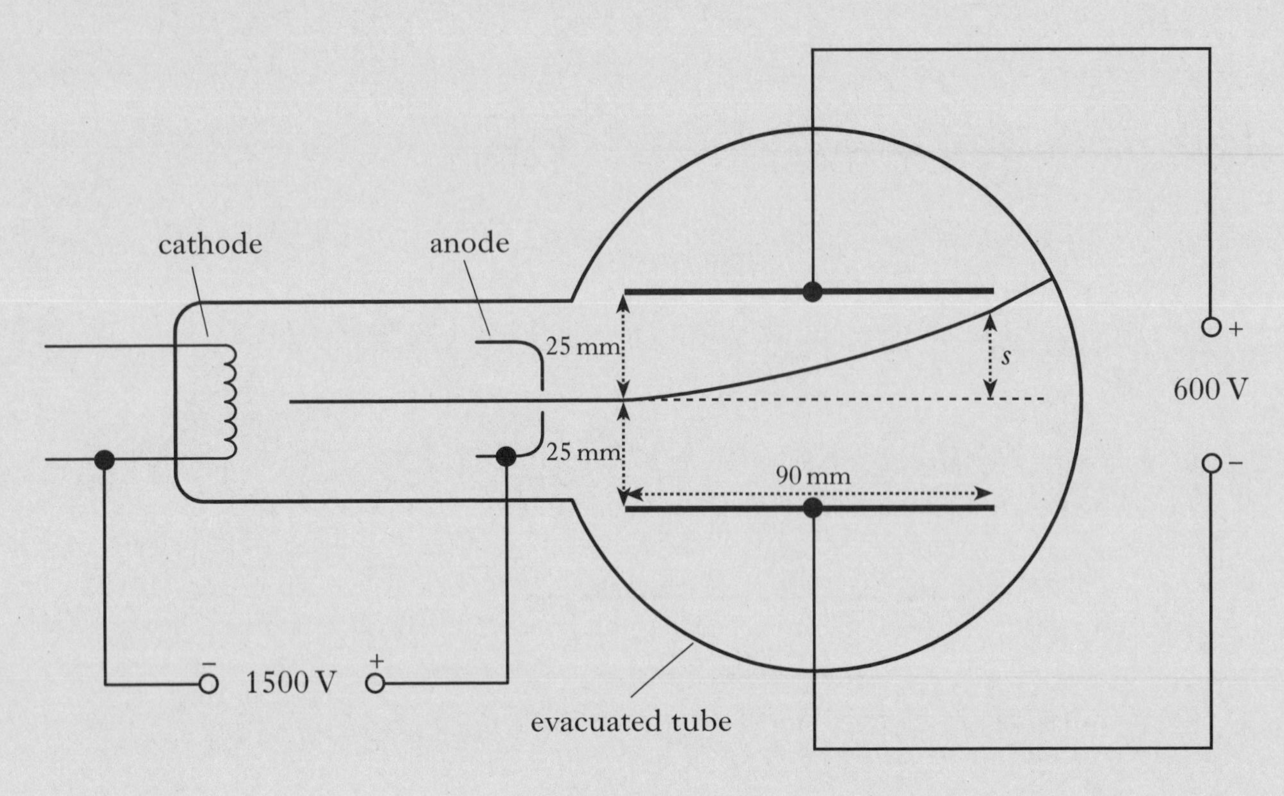

Figure 10

The electrons start from rest at the cathode and are accelerated towards the anode by a potential difference of 1500 V. Electrons enter the electric field at a point midway between the two parallel deflecting plates. The deflecting plates are 90 mm long and 50 mm apart. There is a potential difference of 600 V between the deflecting plates.

(*a*) Show that the speed of an electron at the anode is $2{\cdot}3 \times 10^{7}\,\text{m s}^{-1}$. **2**

(*b*) Calculate the time an electron takes to pass between the deflecting plates. **2**

(*c*) (i) Show that the force which deflects the electron is $1{\cdot}9 \times 10^{-15}$ N. **2**

(ii) Calculate the vertical deflection *s* of an electron as it leaves the space between the deflecting plates. **3**

Marks

6. (continued)

(*d*) Explain, in terms of forces, why the path of the electron is:

(i) curved in the space between the plates; **2**

(ii) straight in the space beyond the plates. **1**

(*e*) The anode voltage is now reduced. State what happens to the value of the vertical deflection *s*. You must justify your answer. **2**

(14)

[Turn over

Marks

7. A charged particle moves with a speed of $2 \cdot 0 \times 10^6 \, m \, s^{-1}$ in a circular orbit in a uniform magnetic field, shown in Figure 11.

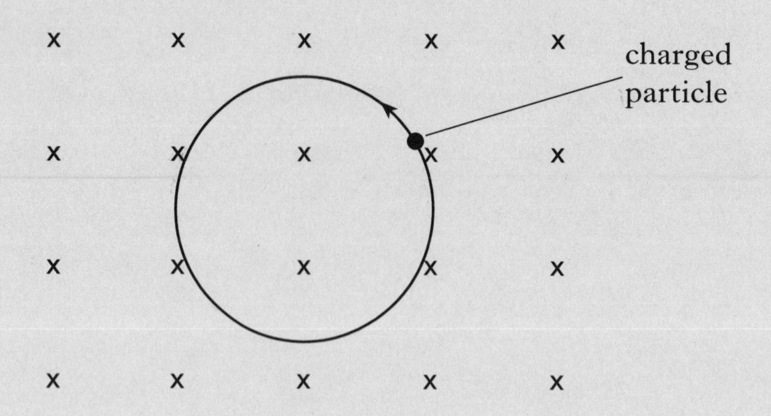

Figure 11

The magnetic induction is $1 \cdot 5$ T and is directed into the page. The circular orbit has a radius of $13 \cdot 9$ mm.

(*a*) (i) State whether the charge on the particle is positive or negative. **1**

(ii) Calculate the charge to mass ratio of the particle. **3**

(iii) Identify the charged particle. You must justify your answer using information from the data sheet. **2**

(*b*) An electron enters a uniform magnetic field at an angle to the magnetic field lines as shown in Figure 12.

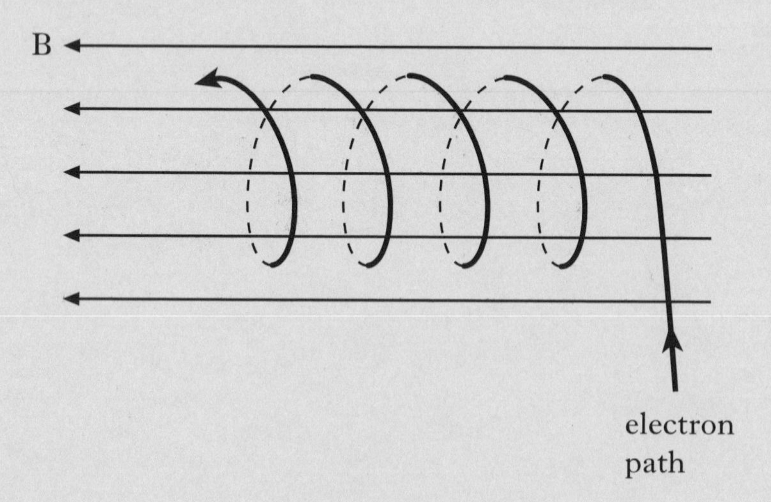

Figure 12

Explain the shape of the electron path in the magnetic field. **2**

Marks

7. (continued)

(*c*) Charged particles which enter the Earth's atmosphere near the North pole collide with air molecules. The light emitted in this process is called the Aurora Borealis.

In Figure 13, the Earth's magnetic field is indicated by continuous lines which show the magnetic field direction in the region surrounding the Earth.

The extent of the Earth's atmosphere is also shown.

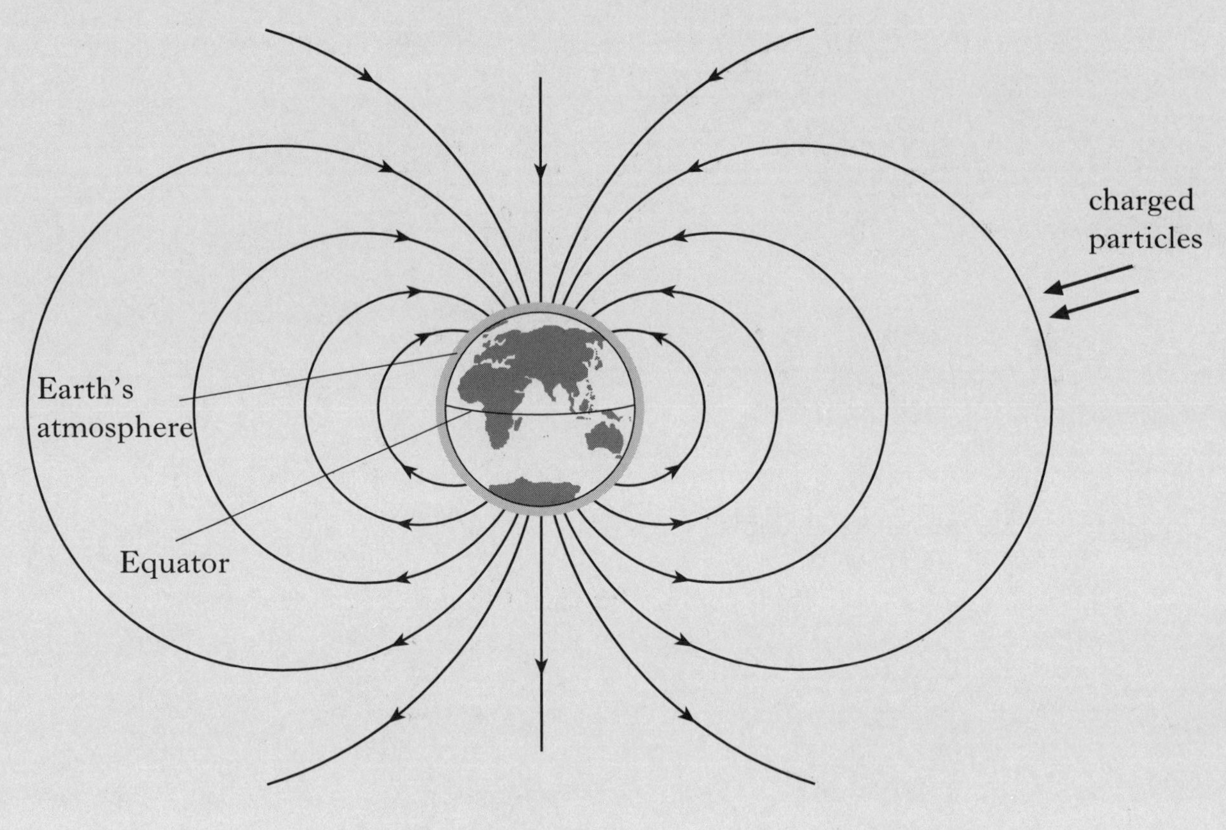

Figure 13

Charged particles approach the Earth in the direction shown in Figure 13. Explain why these particles do **not** cause an aurora above the Equator. **2**

(10)

[Turn over

Marks

8. (*a*) A coil of wire has an inductance of 2·0 H. State what is meant by *an inductance of 2·0 H*. **1**

(*b*) Figure 14 shows a circuit containing an inductor with negligible resistance, a resistor, switch and d.c. power supply connected in series. The d.c. power supply has negligible internal resistance.

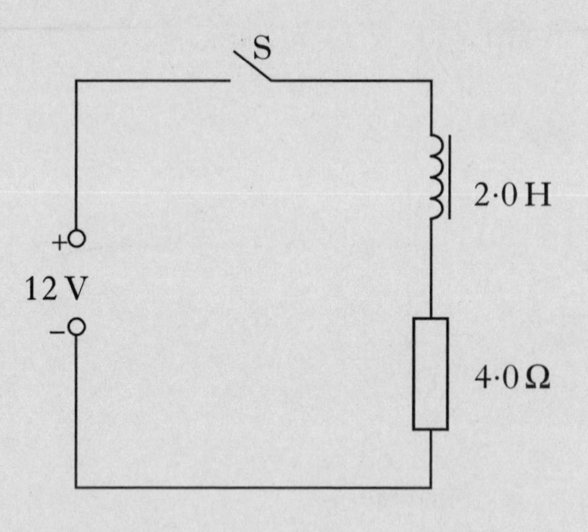

Figure 14

Calculate the rate of change of current immediately after switch S is closed. **2**

(*c*) A similar circuit, with some component values changed, is shown in Figure 15.

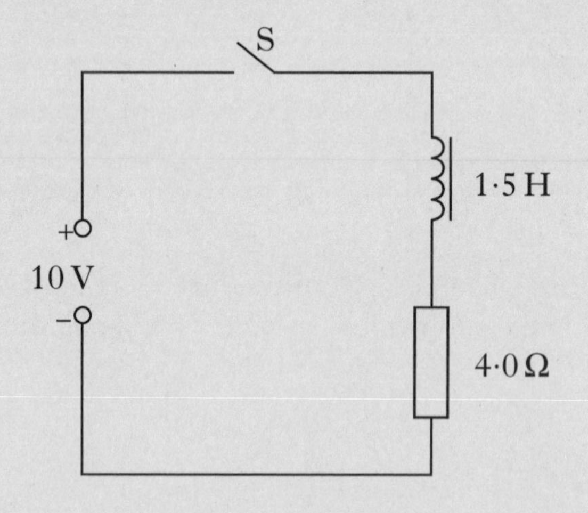

Figure 15

(i) Switch S is closed.

State **two** ways that the current in this circuit differs from the current in the circuit shown in Figure 14. **Justify your answers**. **2**

(ii) Calculate the maximum energy stored in the 1·5 H inductor. **2**

Marks

8. (continued)

(*d*) An airport metal detector consists of two fixed coils as shown in Figure 16.

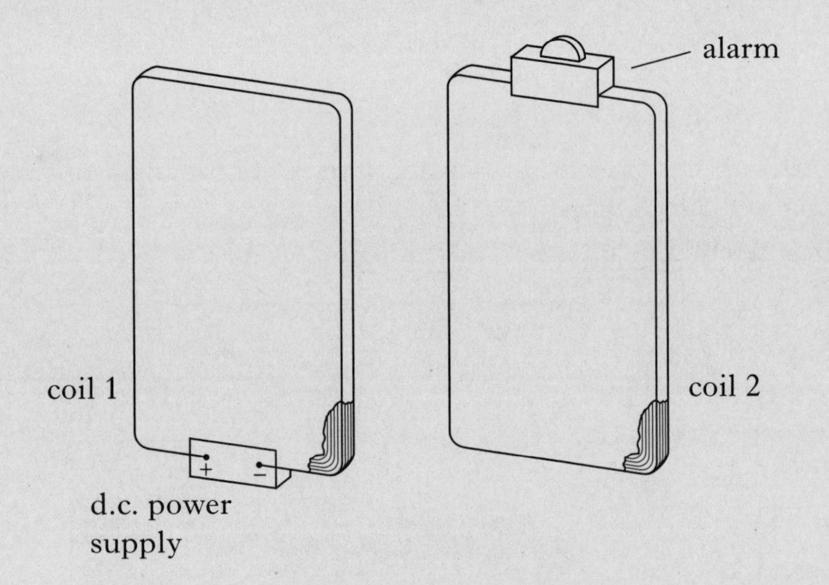

Figure 16

A d.c. power supply provides a current in coil 1.

Coil 2 has no power supply but is connected to an alarm. The alarm triggers when there is a current in coil 2.

A passenger wearing a gold bracelet walks between the coils as shown in Figure 17.

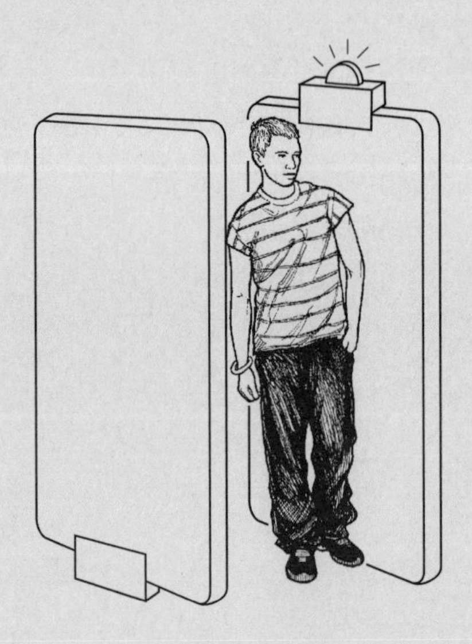

Figure 17

Explain why:

(i) a current is induced in the gold bracelet; **1**

(ii) this triggers the alarm. **1**

(9)

Marks

9. (*a*) A travelling wave is represented by the expression

$$y = 2{\cdot}0 \times 10^{-4} \sin(1570t - 4{\cdot}6x)$$

where x and y are in metres and t is in seconds.

(i) Calculate the frequency of the wave. **2**

(ii) A wave with the same frequency and four times the intensity travels in the opposite direction.

Write down the equation which represents this wave. **2**

(*b*) A train emits a sound of frequency 800 Hz as it passes through a station. The sound is heard by a person on the station platform as shown in Figure 18.

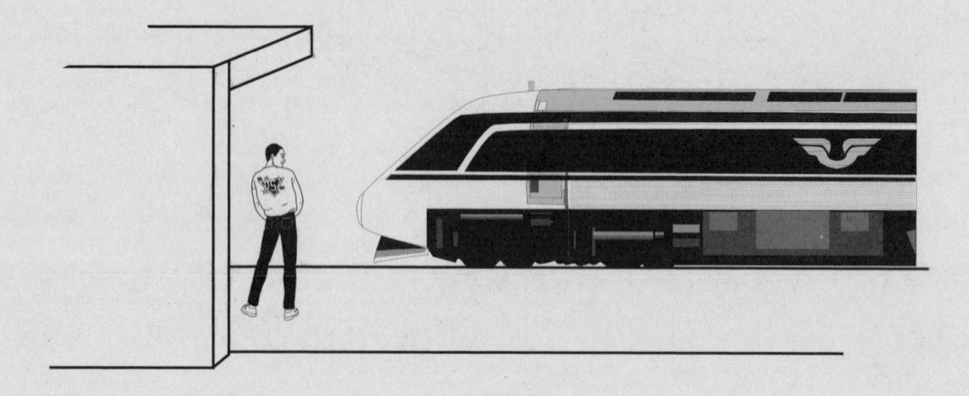

Figure 18

(i) Describe how the frequency of the sound, heard by the person, changes as the train passes through the station. **1**

(ii) Explain, in terms of wavefronts, why this frequency change occurs. You may wish to include a diagram as part of your answer. **2**

(iii) At one instant the person hears a sound of frequency 760 Hz. Calculate the speed of the train relative to the person on the platform at this time. **2**

(9)

Marks

10. A student sets up a Young's slits experiment in order to measure the wavelength of monochromatic light emitted by a laser. The light from the laser passes through a double slit before reaching a screen, where a pattern of light and dark fringes is seen, as shown in Figure 19.

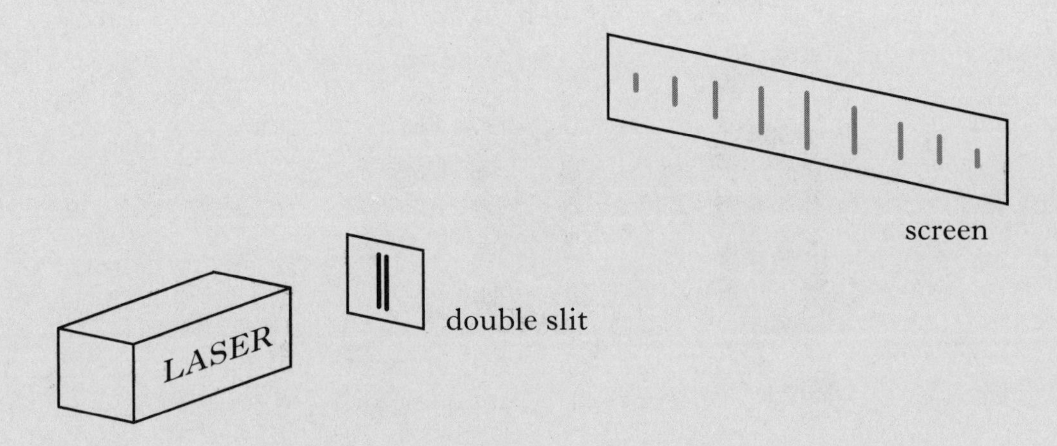

Figure 19

The student records the following measurements:

double slit separation = 0·25 ± 0·01 mm
distance between double slits and screen = 3·91 ± 0·01 m
distance between two adjacent bright fringes = 8·0 ± 0·5 mm.

(*a*) (i) Calculate the wavelength of the laser light. **2**

(ii) Show that the absolute uncertainty in the calculated wavelength is $\pm 4 \times 10^{-8}$ m. **2**

(iii) State why an answer of $\pm 3 \cdot 78 \times 10^{-8}$ m for part (*a*)(ii) would **not** be acceptable. **1**

(*b*) The student now measures the distance between 9 bright fringes (8 spaces). The result is

distance between 9 fringes (8 spaces) = 64·0 ± 0·5 mm.

Calculate the new absolute uncertainty in wavelength, assuming the other measurements remain unchanged. **2**

(*c*) (i) The student then suggests that measuring the distance between 12 bright fringes would significantly reduce the absolute uncertainty in the wavelength. Explain why this is **not** correct. **1**

(ii) State which measurement must be made more accurately to reduce significantly the absolute uncertainty in the wavelength. **1**

(9)

[Turn over

Marks

11. (*a*) State the difference between plane polarised light and unpolarised light. **1**

(*b*) The digital display on a calculator consists of many small segments of liquid crystal material.

A "0" changes to an "8" when the middle segment switches from light to dark as shown in Figure 20.

Figure 20

To make one segment of a 7-segment display, a slice of liquid crystal is placed between a piece of polarising material and a mirror. Figure 21 shows this arrangement for the **middle segment only**.

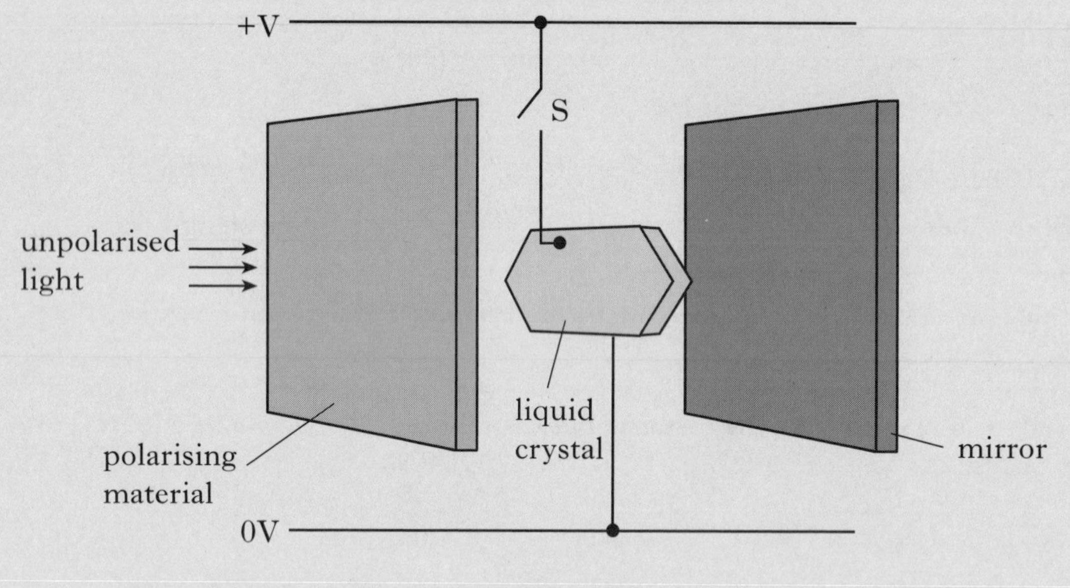

Figure 21

The following table summarises the effect of switch S.

Switch S	*Response of liquid crystal*
open	transmits polarised light
closed	does not transmit polarised light

(i) Explain why the liquid crystal appears dark when switch S is closed. **2**

(ii) State what happens to the switch when an "8" is changed to a "0". **1**

Marks

11. (continued)

(*c*) A student sees a row of numbers displayed on a calculator through a separate piece of polarising material as shown in Figure 22.

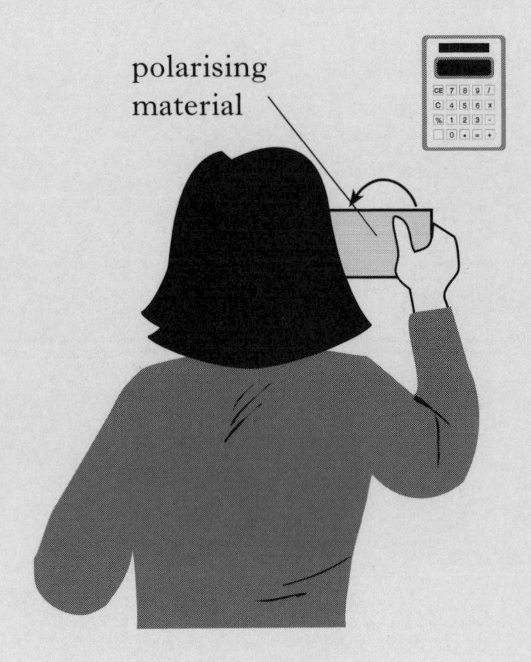

Figure 22

The student rotates the piece of polarising material through 180°. Explain what is seen as the polarising material is rotated. **2**

(6)

[*END OF QUESTION PAPER*]

[BLANK PAGE]

2006 | Advanced Higher

[BLANK PAGE]

X069/701

NATIONAL
QUALIFICATIONS
2006

WEDNESDAY, 17 MAY
1.00 PM – 3.30 PM

PHYSICS
ADVANCED HIGHER

Reference may be made to the Physics Data Booklet.

Answer **all** questions.

Any necessary data may be found in the Data Sheet on page two.

Care should be taken to give an appropriate number of significant figures in the final answers to calculations.

Square-ruled paper (if used) should be placed inside the front cover of the answer book for return to the Scottish Qualifications Authority.

LI X069/701 6/3/4970

DATA SHEET

COMMON PHYSICAL QUANTITIES

Quantity	*Symbol*	*Value*	*Quantity*	*Symbol*	*Value*
Gravitational acceleration on Earth	g	$9{\cdot}8\ \text{m s}^{-2}$	Mass of electron	m_e	$9{\cdot}11 \times 10^{-31}$ kg
Radius of Earth	R_E	$6{\cdot}4 \times 10^{6}$ m	Charge on electron	e	$-1{\cdot}60 \times 10^{-19}$ C
Mass of Earth	M_E	$6{\cdot}0 \times 10^{24}$ kg	Mass of neutron	m_n	$1{\cdot}675 \times 10^{-27}$ kg
Mass of Moon	M_M	$7{\cdot}3 \times 10^{22}$ kg	Mass of proton	m_p	$1{\cdot}673 \times 10^{-27}$ kg
Mean Radius of Moon Orbit		$3{\cdot}84 \times 10^{8}$m	Mass of alpha particle	m_α	$6{\cdot}645 \times 10^{-27}$ kg
Universal constant of gravitation	G	$6{\cdot}67 \times 10^{-11}\ \text{m}^3\ \text{kg}^{-1}\ \text{s}^{-2}$	Charge on alpha particle		$3{\cdot}20 \times 10^{-19}$ C
Speed of light in vacuum	c	$3{\cdot}0 \times 10^{8}\ \text{m s}^{-1}$	Planck's constant	h	$6{\cdot}63 \times 10^{-34}$ J s
Speed of sound in air	v	$3{\cdot}4 \times 10^{2}\ \text{m s}^{-1}$	Permittivity of free space	ε_0	$8{\cdot}85 \times 10^{-12}\ \text{F m}^{-1}$
			Permeability of free space	μ_0	$4\pi \times 10^{-7}\ \text{H m}^{-1}$

REFRACTIVE INDICES

The refractive indices refer to sodium light of wavelength 589 nm and to substances at a temperature of 273 K.

Substance	*Refractive index*	*Substance*	*Refractive index*
Diamond	2·42	Glycerol	1·47
Glass	1·51	Water	1·33
Ice	1·31	Air	1·00
Perspex	1·49	Magnesium Fluoride	1·38

SPECTRAL LINES

Element	*Wavelength*/nm	*Colour*
Hydrogen	656	Red
	486	Blue-green
	434	Blue-violet
	410	Violet
	397	Ultraviolet
	389	Ultraviolet
Sodium	589	Yellow
Cadmium	644	Red
	509	Green
	480	Blue

Lasers

Element	*Wavelength*/nm	*Colour*
Carbon dioxide	9550 } 10590 }	Infrared
Helium-neon	633	Red

PROPERTIES OF SELECTED MATERIALS

Substance	*Density*/ kg m^{-3}	*Melting Point*/ K	*Boiling Point*/ K	*Specific Heat Capacity*/ $\text{J kg}^{-1}\ \text{K}^{-1}$	*Specific Latent Heat of Fusion*/ J kg^{-1}	*Specific Latent Heat of Vaporisation*/ J kg^{-1}
Aluminium	$2{\cdot}70 \times 10^{3}$	933	2623	$9{\cdot}02 \times 10^{2}$	$3{\cdot}95 \times 10^{5}$	....
Copper	$8{\cdot}96 \times 10^{3}$	1357	2853	$3{\cdot}86 \times 10^{2}$	$2{\cdot}05 \times 10^{5}$	....
Glass	$2{\cdot}60 \times 10^{3}$	1400	....	$6{\cdot}70 \times 10^{2}$	....	....
Ice	$9{\cdot}20 \times 10^{2}$	273	....	$2{\cdot}10 \times 10^{3}$	$3{\cdot}34 \times 10^{5}$	....
Glycerol	$1{\cdot}26 \times 10^{3}$	291	563	$2{\cdot}43 \times 10^{3}$	$1{\cdot}81 \times 10^{5}$	$8{\cdot}30 \times 10^{5}$
Methanol	$7{\cdot}91 \times 10^{2}$	175	338	$2{\cdot}52 \times 10^{3}$	$9{\cdot}9 \times 10^{4}$	$1{\cdot}12 \times 10^{6}$
Sea Water	$1{\cdot}02 \times 10^{3}$	264	377	$3{\cdot}93 \times 10^{3}$	....	....
Water	$1{\cdot}00 \times 10^{3}$	273	373	$4{\cdot}19 \times 10^{3}$	$3{\cdot}34 \times 10^{5}$	$2{\cdot}26 \times 10^{6}$
Air	1·29	....	....	....	....	....
Hydrogen	$9{\cdot}0 \times 10^{-2}$	14	20	$1{\cdot}43 \times 10^{4}$	....	$4{\cdot}50 \times 10^{5}$
Nitrogen	1·25	63	77	$1{\cdot}04 \times 10^{3}$	....	$2{\cdot}00 \times 10^{5}$
Oxygen	1·43	55	90	$9{\cdot}18 \times 10^{2}$	....	$2{\cdot}40 \times 10^{5}$

The gas densities refer to a temperature of 273 K and a pressure of $1{\cdot}01 \times 10^{5}$ Pa.

[BLANK PAGE]

[Turn over for Question 1 on *Page four*

Marks

1. A child's toy consists of a model aircraft attached to a light cord. The aircraft is swung in a vertical circle **at constant speed** as shown in Figure 1.

X is the highest point and Y the lowest point in the circle.

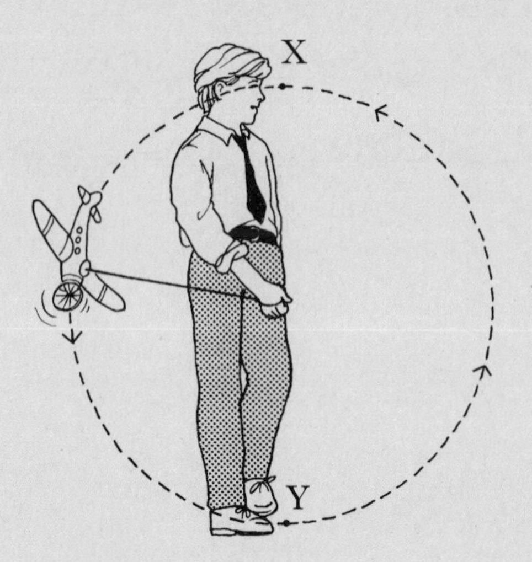

Figure 1

(*a*) The time taken for the aircraft to complete 20 revolutions is measured five times.

The mass of the aircraft and the radius of the circle are also measured. The following data is obtained.

Time for 20 revolutions: 10·05 s; 9·88 s; 10·30 s; 9·80 s; 9·97 s.
Radius of circle = 0·500 ± 0·002 m.
Mass of aircraft = 0·200 ± 0·008 kg.

(i) (A) Calculate the average period of revolution of the aircraft.

(B) Assuming that the scale reading uncertainty and the calibration uncertainty of the timer are negligible, calculate the absolute uncertainty in the period. **3**

(ii) Show that the centripetal force acting on the aircraft is 15·8 N. **2**

(iii) Calculate the absolute uncertainty in this value for the centripetal force. Express your answer in the form

$$F = (15{\cdot}8 \pm \qquad) \text{ N}.$$ **4**

(iv) Draw labelled diagrams to show the forces acting on the aircraft:

(A) at position X;

(B) at position Y. **2**

(v) Calculate the minimum tension in the cord. **2**

Marks

1. (continued)

(*b*) The aircraft has a small air siren which produces a note of frequency 1000 Hz.

One student swings the aircraft in a vertical circle at constant speed. A second student listens to the note while standing in front of the first student, as shown in Figure 2.

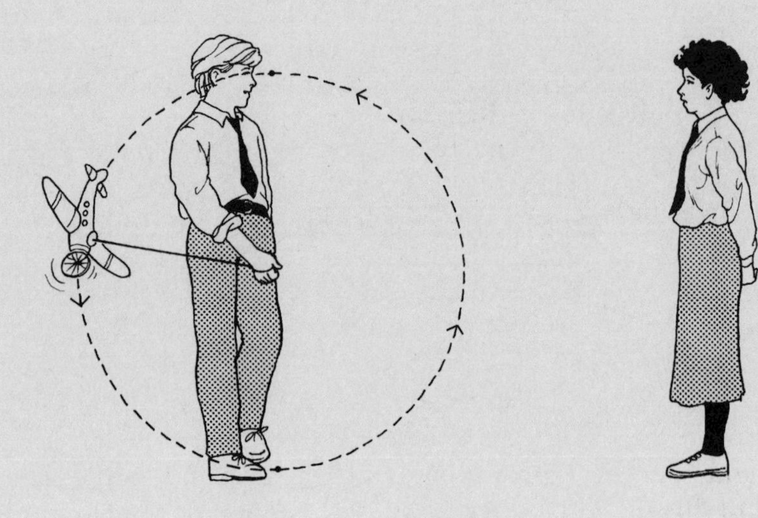

Figure 2

State what happens to the pitch of the note heard by the second student as the aircraft passes through its highest point. **1**

(14)

[Turn over

Marks

2. A circular metal disc is mounted horizontally on the axle of a rotational motion sensor as shown in Figure 3.

The axle is on a frictionless bearing.

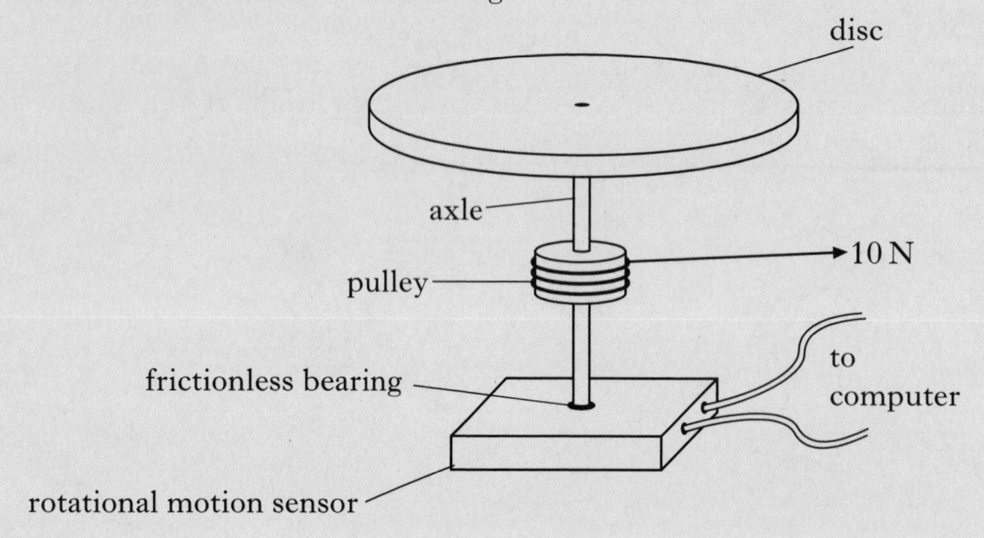

Figure 3

A thin cord is wound round a light pulley which is attached to the axle. The pulley has a radius of 20 mm and a force of 10 N is applied to the free end of the cord.

The cord fully unwinds from the pulley in a time of 3·0 s.

The rotational motion sensor is interfaced to a computer which is programmed to display a graph showing the variation of the angular velocity of the metal disc with time.

The graph displayed on the monitor is shown in Figure 4.

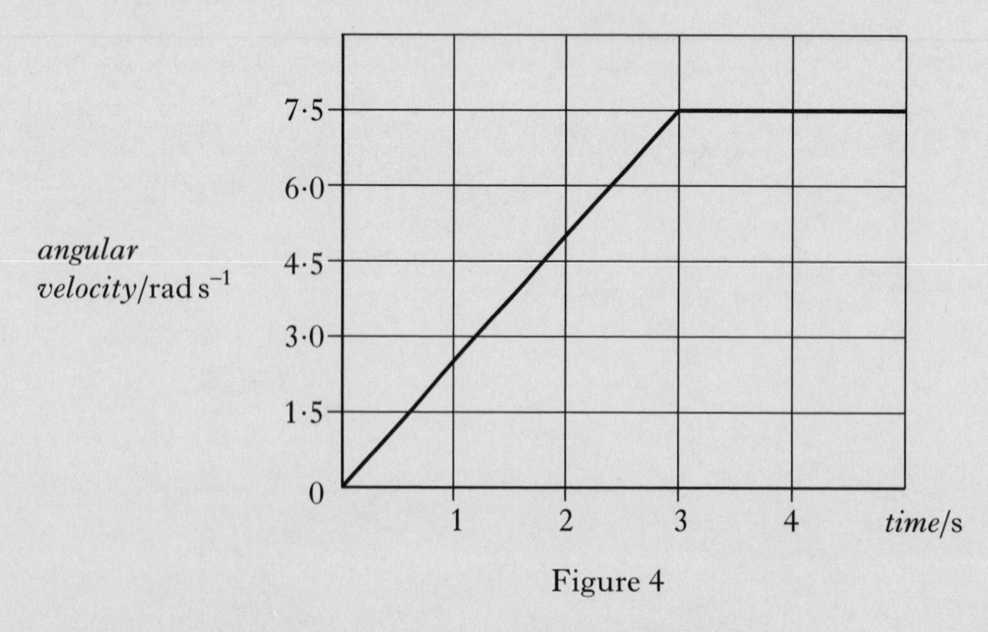

Figure 4

(*a*) (i) Calculate the torque exerted by the cord. **2**

(ii) Using information from the graph, determine the angular acceleration of the disc. **2**

(iii) Calculate the moment of inertia of the disc. **2**

Marks

2. (continued)

(*b*) After the cord is fully unwound, a second uniform disc with mass 3·2 kg and radius 0·12 m is gently dropped on top of the original disc as shown in Figure 5.

Both discs now rotate with a new angular velocity.

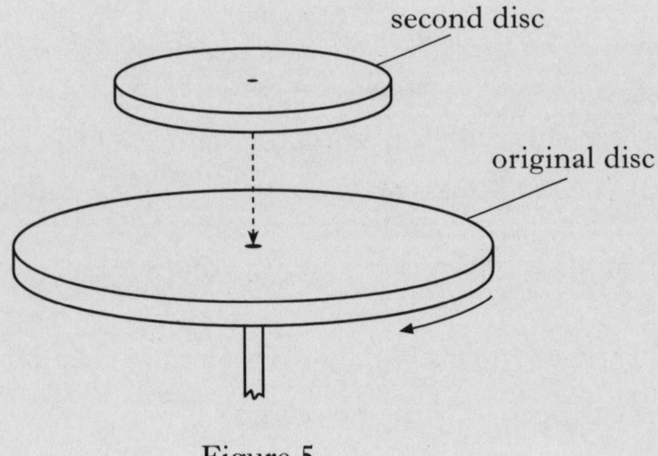

Figure 5

(i) Calculate the moment of inertia of the second disc. **2**

(ii) Calculate the new angular velocity of the discs. **2**

(*c*) The experiment is repeated, except that a **ring**, with the same mass and diameter as the second disc, is gently dropped on top of the original disc as shown in Figure 6.

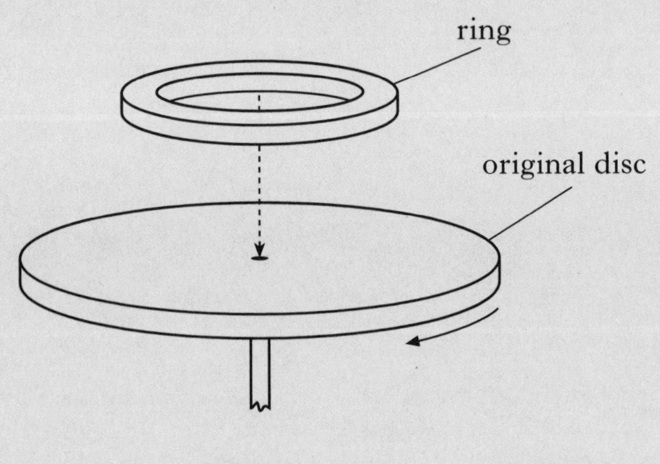

Figure 6

State whether the resulting angular velocity is greater than, less than or the same as that calculated in (*b*)(ii).

You must justify your answer. **2**

(12)

[Turn over

Marks

3. (*a*) (i) State what is meant by *gravitational field strength*. **1**

(ii) The gravitational field strength at the surface of Mars is $3 \cdot 7 \, \mathrm{N\,kg^{-1}}$.

The radius of Mars is $3 \cdot 4 \times 10^3 \, \mathrm{km}$.

(A) Use Newton's universal law of gravitation to show that the mass of Mars is given by the equation

$$M = \frac{gr^2}{G}$$

where the symbols have their usual meaning.

(B) Calculate the mass of Mars. **3**

(*b*) A spacecraft of mass 100 kg is in circular orbit 300 km above the surface of Mars.

(i) Show that the force exerted by Mars on the spacecraft is $3 \cdot 1 \times 10^2 \, \mathrm{N}$. **2**

(ii) Calculate the period of the spacecraft's orbit. **3**

(9)

Marks

4. A test tube contains lead shot. The combined mass of the test tube and the lead shot is 0·250 kg.

The test tube is gently dropped into a container of water and oscillates above and below its equilibrium position with simple harmonic motion as shown in Figure 7.

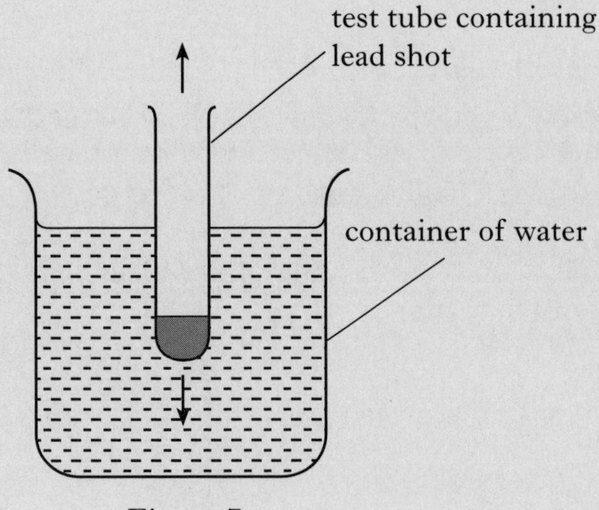

Figure 7

The displacement y of the test tube from its equilibrium position is described by the equation

$$y = 0{\cdot}05 \cos 6t$$

where y is in metres and t is in seconds.

(*a*) Show that the kinetic energy of the test tube, in joules, is given by the equation

$$E_k = 4{\cdot}5\ (2{\cdot}5 \times 10^{-3} - y^2).$$ **2**

(*b*) Calculate the maximum value of the kinetic energy of the test tube. **1**

(*c*) Calculate the potential energy of the test tube when it is 40 mm above its equilibrium position. **2**

(5)

[Turn over

Marks

5. (*a*) A charged metal sphere has a diameter of 0·36 m. The electrostatic potential at the surface of the sphere is $+2{\cdot}8 \times 10^5$ V.

(i) Show that the charge on the sphere is $+5{\cdot}6 \times 10^{-6}$ C. **2**

(ii) State the electrostatic potential at a point 0·10 m from the **centre** of the sphere. **1**

(iii) (A) Calculate the electric field strength at the surface of the sphere.

(B) Sketch a graph of the electric field strength against distance from the centre of the sphere to a point well beyond the sphere's surface. No numerical values are required. **3**

(*b*) Two identical spheres, each carrying a charge of $+5{\cdot}6 \times 10^{-6}$ C, are now placed as shown in Figure 8.

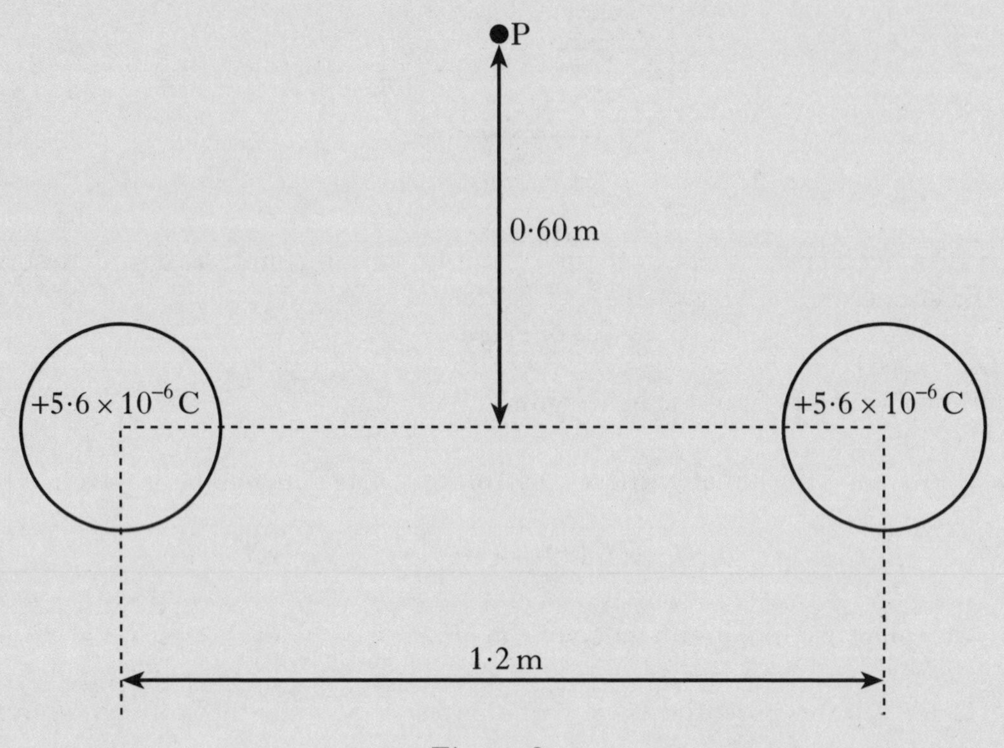

Figure 8

Point P is 0·60 m vertically above the mid-point of the line joining the centres of the two spheres.

Determine the magnitude and direction of the electric field strength at point P. **4**

(10)

Marks

6. The print head of an ink-jet printer fires a tiny drop of ink of mass $1{\cdot}2 \times 10^{-12}$ kg as it scans across the paper. The drop carries a charge of $-1{\cdot}6 \times 10^{-12}$ C and enters the space between a pair of parallel plates at a speed of $20\,\mathrm{m\,s^{-1}}$, as shown in Figure 9.

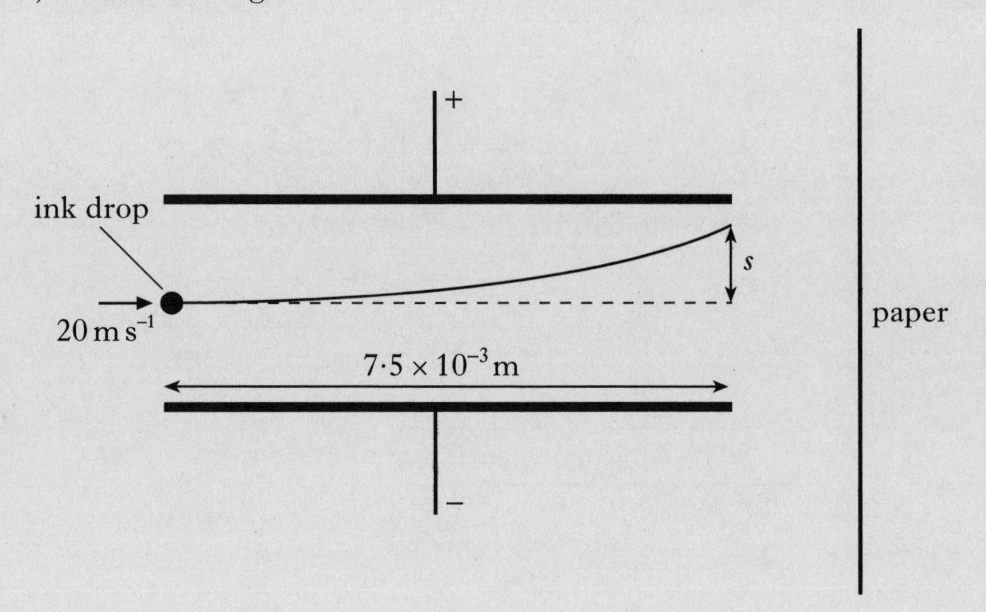

Figure 9

The length of the plates is $7{\cdot}5 \times 10^{-3}$ m and the electric field strength between them is $2{\cdot}5 \times 10^{4}\,\mathrm{N\,C^{-1}}$.

(*a*) Calculate the magnitude of the electrostatic force acting on the ink drop as it passes between the plates. **2**

(*b*) Show, by calculation, that the gravitational force acting on the drop is negligible compared to the electrostatic force. **2**

(*c*) Calculate the deflection *s* of the drop as it leaves the region between the plates. **4**

(*d*) Calculate the number of excess electrons on the ink drop. **2**

(10)

[Turn over

Marks

7. (*a*) (i) A cyclotron is a particle accelerator which consists of two D-shaped hollow structures, called "dees", placed in a vacuum. Figure 10 shows an arrangement for a cyclotron.

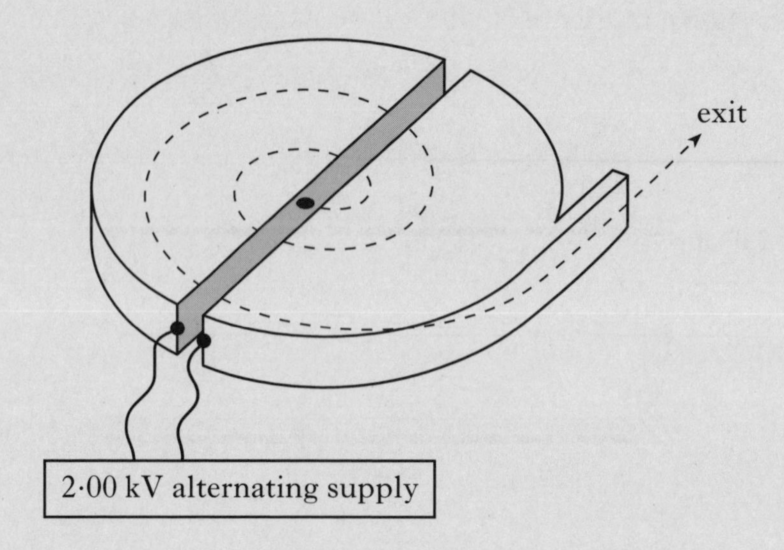

Figure 10

Figure 11 shows the cyclotron viewed from above.

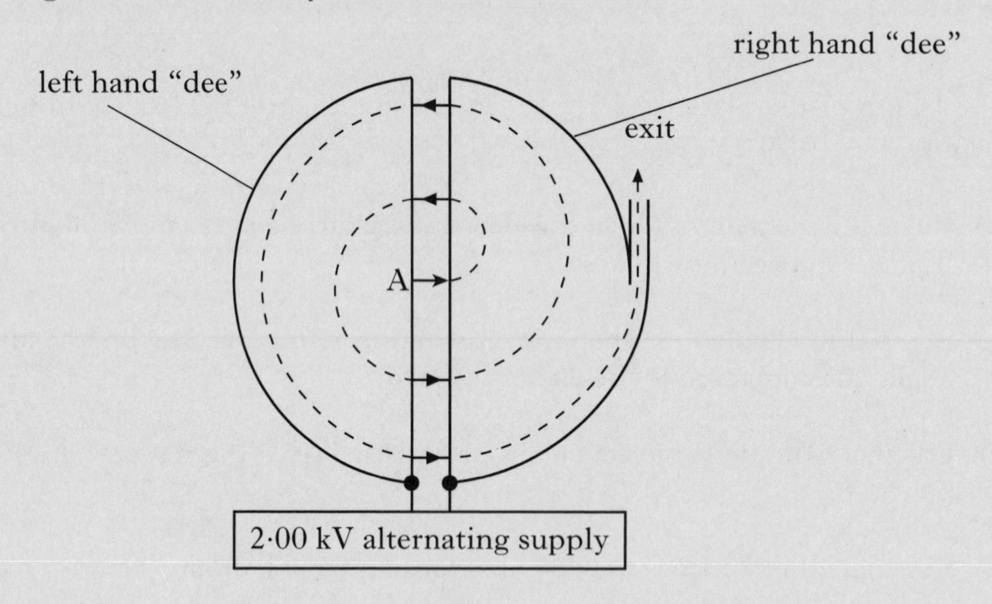

Figure 11

Protons are released from rest at point A and accelerated across the gap between the "dees" by a voltage of 2·00 kV.

Show that the speed of the protons as they **first** reach the right hand "dee" is $6{\cdot}19 \times 10^5\,\text{m s}^{-1}$. **2**

(ii) Inside the "dees" the electric field strength is zero but a uniform magnetic field of 1·30 T acts perpendicularly to the "dees". This forces the protons to move in semi-circular paths when inside the "dees".

Calculate the radius of the **first** such path in the right hand "dee". **3**

Marks

7. (*a*) (continued)

(iii) While the protons are inside the "dee", the polarity of the applied voltage is reversed so that the protons are again accelerated when they cross to the left hand "dee".

Calculate the speed of the protons as they **first** enter the left hand "dee". **2**

(*b*) The protons exit the cyclotron with a kinetic energy of $1 \cdot 57 \times 10^{-13}$ J and are aimed at a gold target. The charge on a gold nucleus is $+79e$.

Find the distance of closest approach to a gold nucleus for a proton from this cyclotron. **3**

(*c*) A larger accelerator produces protons with a **relativistic mass** of $4 \cdot 66 \times 10^{-27}$ kg.

Calculate the speed of these protons. **2**

(12)

[Turn over

Marks

8. A datalogger is used to investigate the rate of change of current in the circuit shown in Figure 12.

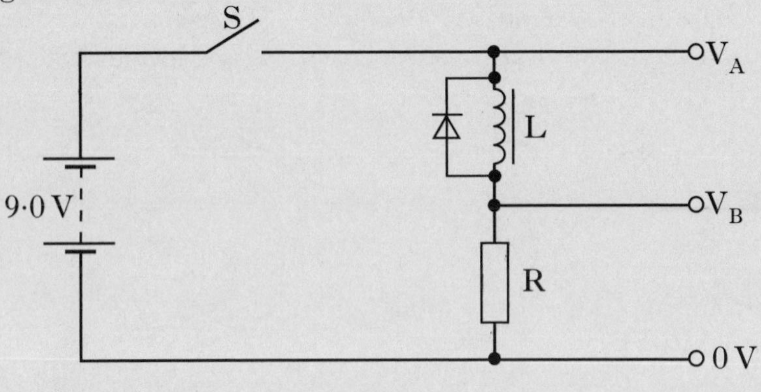

Figure 12

(*a*) The datalogger measures the potential V_A and the potential V_B.

What other piece of information is required to allow the computer software to determine the current in the circuit? **1**

(*b*) The switch S is closed and the datalogger software produces the graph shown in Figure 13.

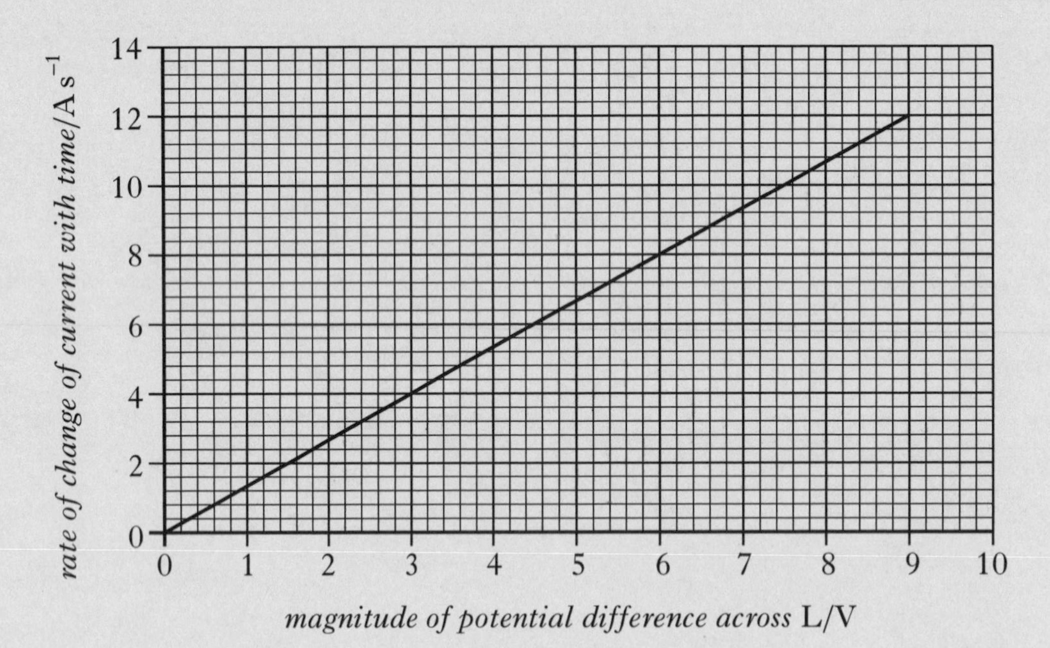

Figure 13

Assuming that the resistance of the inductor is negligible, calculate its inductance. **2**

(*c*) The current in the circuit eventually reaches a steady value of 100 mA.

Calculate the energy stored in the magnetic field of the inductor. **2**

(*d*) The diode in the circuit is necessary to protect the datalogger against the high voltage which can arise when the switch S is opened.

Explain why this high voltage is produced. **1**

(6)

Marks

9. (*a*) According to modern particle theory, protons and neutrons are composed of combinations of up and down quarks. Up quarks have a charge of $+\frac{2}{3}e$ while down quarks have a charge of $-\frac{1}{3}e$.

(i) Name the force which holds the quarks together in protons and neutrons. **1**

(ii) State the combination of up and down quarks which make up:

(A) a proton;

(B) a neutron. **2**

(*b*) A neutron can decay into a proton, electron and antineutrino.

Name the force associated with this decay. **1**

(*c*) A thermal neutron has a velocity of $3{\cdot}5 \times 10^3\,\mathrm{m\,s^{-1}}$.

Calculate the de Broglie wavelength of this neutron. **2**

(6)

[Turn over

Marks

10. In an experiment to measure the speed of sound in air, a loudspeaker, a signal generator and a reflector are set up as shown in Figure 14.

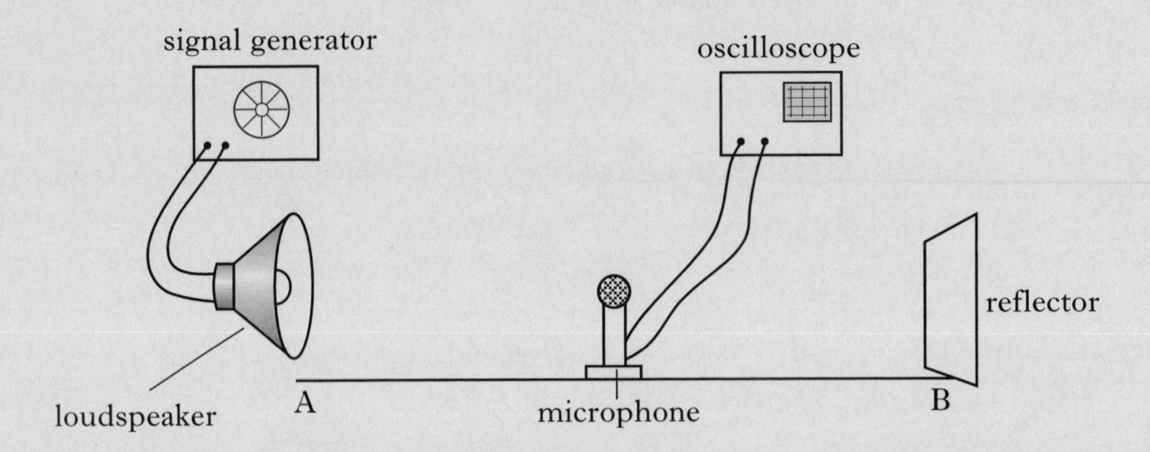

Figure 14

A stationary wave pattern is produced between the loudspeaker and the reflector. The intensity of the sound is monitored using a microphone connected to an oscilloscope. The microphone is moved steadily along line AB and the oscilloscope indicates alternate maximum and minimum values of sound intensity.

(*a*) What name is given to the points in the stationary wave pattern at which **minimum** values of sound intensity occur? **1**

(*b*) The signal generator is adjusted until the frequency of the sound produced is 2000 Hz. The distance between two successive points of minimum sound intensity is measured as 88 mm.

(i) Use this data to calculate the speed of sound in air. **3**

(ii) Suggest **one** improvement to the experiment which would result in a more accurate value for the speed of sound in air.

Justify your answer. **2**

(*c*) The microphone is placed at a position of minimum sound intensity. Without moving the microphone, the reflector is moved away from the loudspeaker until a minimum is again detected.

The intensity of sound at this minimum is found to be **greater** than the intensity of sound before the reflector was moved.

Explain this observation. **2**

(8)

Marks

11. (*a*) An air wedge is formed between two flat glass plates of length l, which are in contact at one end. They are separated by a human hair of diameter d at the other end, as shown in Figure 15.

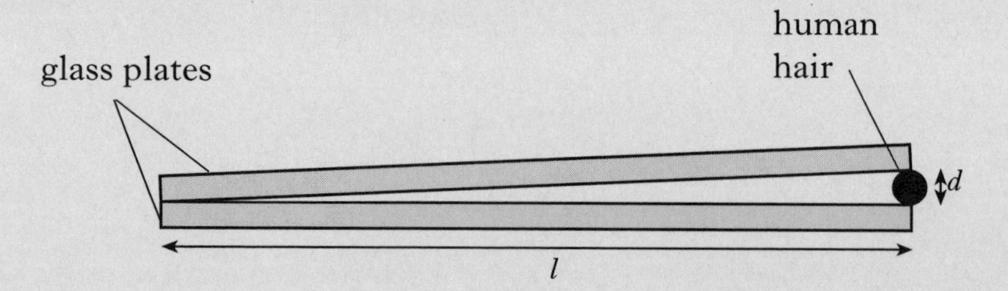

Figure 15

The air wedge is illuminated from above by a monochromatic light source of wavelength λ. When viewed from above a series of interference fringes of separation Δx is observed.

(i) Use this information to derive an expression for the diameter of the human hair. **2**

(ii) The wavelength of the monochromatic light is 589 nm, the length of the glass plates is 75 mm and the separation between two adjacent dark fringes is $3{\cdot}4 \times 10^{-4}$ m.

Calculate the diameter of the hair. **1**

(*b*) A camera lens can be made non-reflecting by coating it with a thin layer of magnesium fluoride.

(i) Calculate the thickness of magnesium fluoride required to make the lens non-reflecting for light of wavelength 548 nm. **2**

(ii) The lens has a thin film of transparent liquid placed on its surface as shown in Figure 16. The refractive index of the liquid is 1·45.

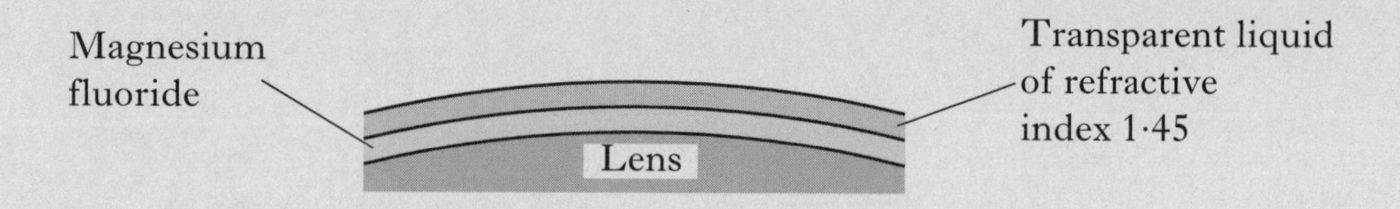

Figure 16

Explain why the coating is no longer non-reflective. **2**

(*c*) Explain why coloured fringes can be observed when a thin film of oil forms on a puddle of water. **1**

(8)

[*END OF QUESTION PAPER*]

[BLANK PAGE]

2007 | Advanced Higher

[BLANK PAGE]

X069/701

NATIONAL QUALIFICATIONS 2007

WEDNESDAY, 16 MAY
1.00 PM – 3.30 PM

PHYSICS
ADVANCED HIGHER

Reference may be made to the Physics Data Booklet.

Answer **all** questions.

Any necessary data may be found in the Data Sheet on page two.

Care should be taken to give an appropriate number of significant figures in the final answers to calculations.

Square-ruled paper (if used) should be placed inside the front cover of the answer book for return to the Scottish Qualifications Authority.

DATA SHEET

COMMON PHYSICAL QUANTITIES

Quantity	*Symbol*	*Value*	*Quantity*	*Symbol*	*Value*
Gravitational acceleration on Earth	g	$9\cdot8$ m s^{-2}	Mass of electron	m_e	$9\cdot11 \times 10^{-31}$ kg
Radius of Earth	R_E	$6\cdot4 \times 10^6$ m	Charge on electron	e	$-1\cdot60 \times 10^{-19}$ C
Mass of Earth	M_E	$6\cdot0 \times 10^{24}$ kg	Mass of neutron	m_n	$1\cdot675 \times 10^{-27}$ kg
Mass of Moon	M_M	$7\cdot3 \times 10^{22}$ kg	Mass of proton	m_p	$1\cdot673 \times 10^{-27}$ kg
Radius of Moon	R_M	$1\cdot7 \times 10^6$ m	Mass of alpha particle	m_α	$6\cdot645 \times 10^{-27}$ kg
Mean Radius of Moon Orbit		$3\cdot84 \times 10^8$ m	Charge on alpha particle		$3\cdot20 \times 10^{-19}$ C
Universal constant of gravitation	G	$6\cdot67 \times 10^{-11}$ m^3 kg^{-1} s^{-2}	Planck's constant	h	$6\cdot63 \times 10^{-34}$ J s
Speed of light in vacuum	c	$3\cdot0 \times 10^8$ m s^{-1}	Permittivity of free space	ε_0	$8\cdot85 \times 10^{-12}$ F m^{-1}
Speed of sound in air	v	$3\cdot4 \times 10^2$ m s^{-1}	Permeability of free space	μ_0	$4\pi \times 10^{-7}$ H m^{-1}

REFRACTIVE INDICES

The refractive indices refer to sodium light of wavelength 589 nm and to substances at a temperature of 273 K.

Substance	*Refractive index*	*Substance*	*Refractive index*
Diamond	2·42	Glycerol	1·47
Glass	1·51	Water	1·33
Ice	1·31	Air	1·00
Perspex	1·49	Magnesium Fluoride	1·38

SPECTRAL LINES

Element	*Wavelength*/nm	*Colour*
Hydrogen	656	Red
	486	Blue-green
	434	Blue-violet
	410	Violet
	397	Ultraviolet
	389	Ultraviolet
Sodium	589	Yellow

Element	*Wavelength*/nm	*Colour*
Cadmium	644	Red
	509	Green
	480	Blue

Lasers

Element	*Wavelength*/nm	*Colour*
Carbon dioxide	9550 } 10590 }	Infrared
Helium-neon	633	Red

PROPERTIES OF SELECTED MATERIALS

Substance	*Density*/ kg m^{-3}	*Melting Point*/ K	*Boiling Point*/ K	*Specific Heat Capacity*/ J kg^{-1} K^{-1}	*Specific Latent Heat of Fusion*/ J kg^{-1}	*Specific Latent Heat of Vaporisation*/ J kg^{-1}
Aluminium	$2\cdot70 \times 10^3$	933	2623	$9\cdot02 \times 10^2$	$3\cdot95 \times 10^5$	....
Copper	$8\cdot96 \times 10^3$	1357	2853	$3\cdot86 \times 10^2$	$2\cdot05 \times 10^5$	....
Glass	$2\cdot60 \times 10^3$	1400	....	$6\cdot70 \times 10^2$	....	....
Ice	$9\cdot20 \times 10^2$	273	....	$2\cdot10 \times 10^3$	$3\cdot34 \times 10^5$	....
Glycerol	$1\cdot26 \times 10^3$	291	563	$2\cdot43 \times 10^3$	$1\cdot81 \times 10^5$	$8\cdot30 \times 10^5$
Methanol	$7\cdot91 \times 10^2$	175	338	$2\cdot52 \times 10^3$	$9\cdot9 \times 10^4$	$1\cdot12 \times 10^6$
Sea Water	$1\cdot02 \times 10^3$	264	377	$3\cdot93 \times 10^3$	....	....
Water	$1\cdot00 \times 10^3$	273	373	$4\cdot19 \times 10^3$	$3\cdot34 \times 10^5$	$2\cdot26 \times 10^6$
Air	1·29	....	....	....	....	....
Hydrogen	$9\cdot0 \times 10^{-2}$	14	20	$1\cdot43 \times 10^4$	....	$4\cdot50 \times 10^5$
Nitrogen	1·25	63	77	$1\cdot04 \times 10^3$	....	$2\cdot00 \times 10^5$
Oxygen	1·43	55	90	$9\cdot18 \times 10^2$	....	$2\cdot40 \times 10^5$

The gas densities refer to a temperature of 273 K and a pressure of $1\cdot01 \times 10^5$ Pa.

Marks

1. (*a*) A particle has displacement $s = 0$ at time $t = 0$ and moves with constant acceleration a.

The velocity of the object is given by the equation $v = u + at$, where the symbols have their usual meanings.

Using calculus, derive an equation for the displacement s of the object as a function of time t. **2**

(*b*) A cyclotron accelerates protons to a velocity of $2{\cdot}80 \times 10^8\,\mathrm{m\,s^{-1}}$.

Calculate the relativistic energy of a proton at this velocity. **4**

(6)

[Turn over

Marks

2. (*a*) A turntable consists of a uniform disc of radius 0·15 m and mass 0·60 kg.

(i) Calculate the moment of inertia of the turntable about the axis of rotation shown in Figure 1. **2**

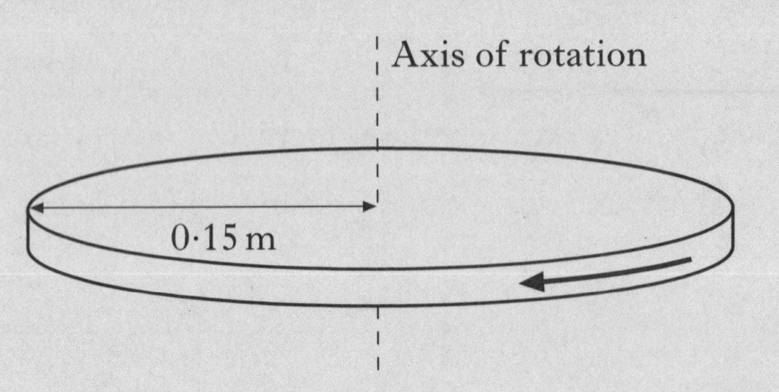

Figure 1

(ii) The turntable accelerates uniformly from rest until it rotates at 45 revolutions per minute. The time taken for the acceleration is 1·5 s.

(A) Show that the angular velocity after 1·5 s is 4·7 rad s^{-1}. **1**

(B) Calculate the angular acceleration of the turntable. **2**

(iii) When the turntable is rotating at 45 revolutions per minute, its motor is disengaged. The turntable continues to rotate freely with negligible friction.

A small mass of 0·20 kg is dropped onto the turntable at a distance of 0·10 m from the centre, as shown in Figure 2.

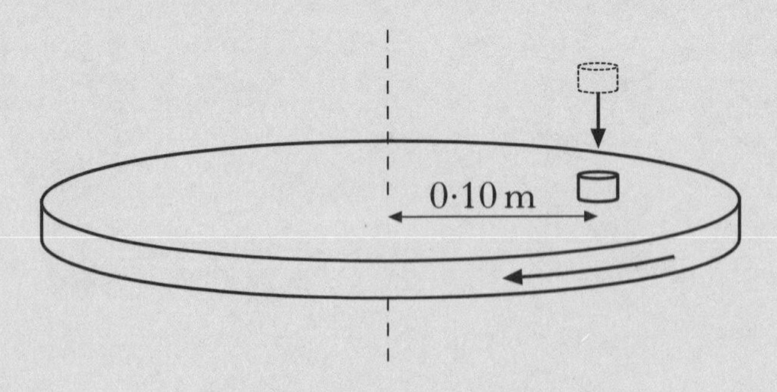

Figure 2

The mass remains in this position on the turntable due to friction, and the turntable and mass rotate together.

Calculate the new angular velocity of the turntable and mass. **3**

Marks

2. (*a*) (continued)

(iv) The experiment is repeated, but the mass is dropped at a distance greater than 0·10 m from the centre of the turntable. The mass slides off the turntable.

Explain why this happens. **2**

(*b*) An ice-skater spins with her arms and one leg outstretched as shown in Figure 3(*a*). She then pulls her arms and leg close to her body as shown in Figure 3(*b*).

Figure 3(*a*)

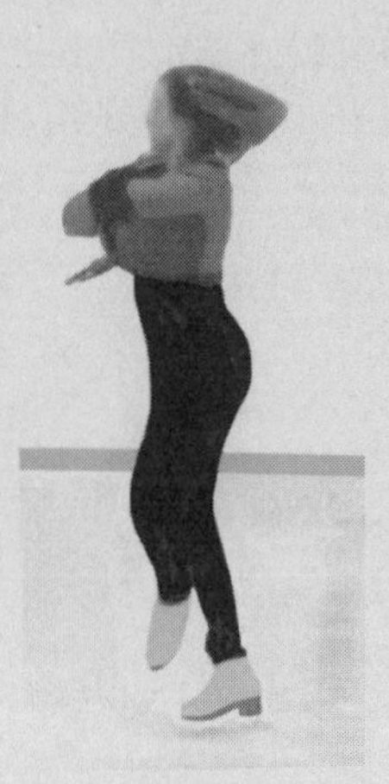

Figure 3(*b*)

State what happens to her angular velocity during this manoeuvre.

Justify your answer. **2**

(12)

[Turn over

Marks

3. (*a*) The Moon orbits the Earth due to the gravitational force between them.

(i) Calculate the magnitude of the gravitational force between the Earth and the Moon. **2**

(ii) Hence calculate the tangential speed of the Moon in its orbit around the Earth. **2**

(iii) Define the term *gravitational potential* at a point in space. **1**

(iv) Calculate the potential energy of the Moon in its orbit. **2**

(v) Hence calculate the total energy of the Moon in its orbit. **2**

(*b*) (i) Derive an expression for the escape velocity from the surface of an astronomical body. **2**

(ii) Calculate the escape velocity from the surface of the Moon. **2**

(13)

Marks

4. (*a*) State what is meant by *simple harmonic motion.* **1**

(*b*) The motion of a piston in a car engine closely approximates to simple harmonic motion.

In a typical engine, the top of a piston moves up and down between points A and B, a distance of 0·10 m, as shown in Figure 4.

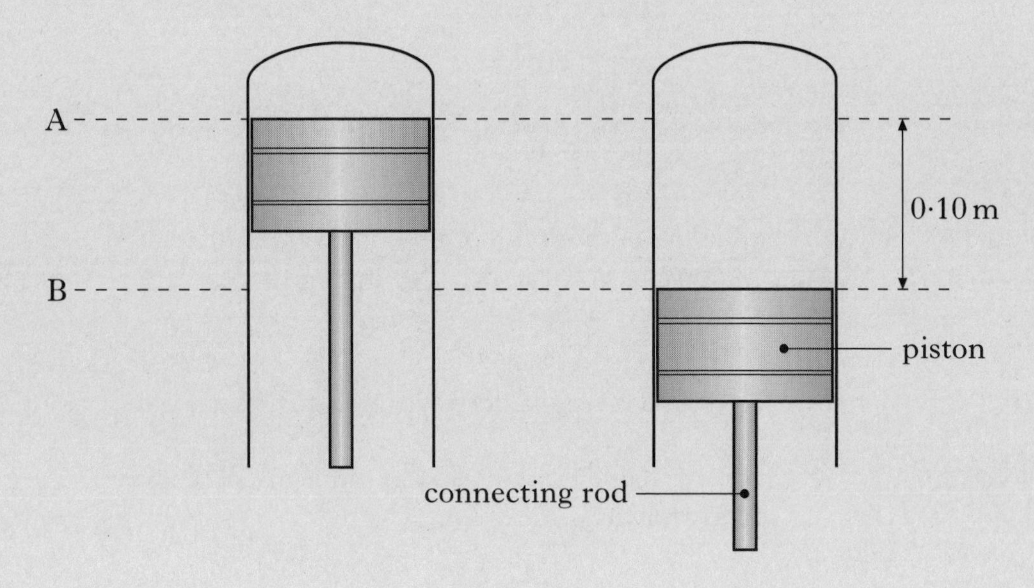

Figure 4

The frequency of the piston's motion is 100 Hz.

Write down an equation which describes how the displacement of the piston from its central position varies with time. Numerical values are required. **2**

(*c*) Calculate the maximum acceleration of the piston. **2**

(*d*) The mass of the piston is 0·48 kg.

Calculate the maximum force applied to the piston by the connecting rod. **2**

(*e*) Calculate the maximum kinetic energy of the piston. **2**

(9)

[Turn over

Marks

5. (*a*) Figure 5 shows a point charge of +5·1 nC.

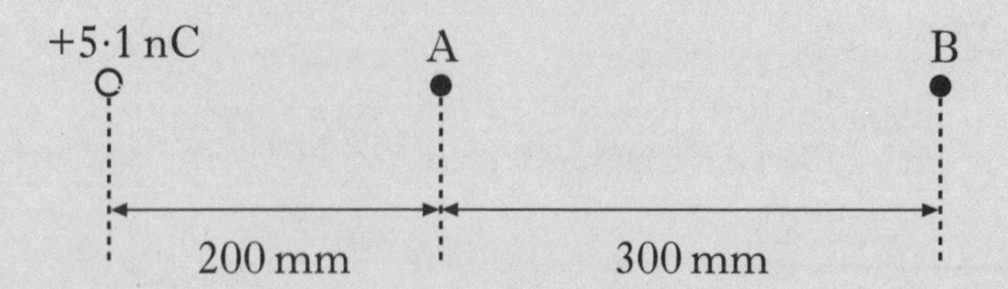

Figure 5

Point A is a distance of 200 mm from the point charge.
Point B is a distance of 300 mm from point A as shown in Figure 5.

(i) Show that the potential at point A is 230 V. **1**

(ii) Calculate the potential difference between A and B. **2**

(*b*) A conducting sphere on an insulating support is some distance away from a negatively charged rod as shown in Figure 6.

Figure 6

Using diagrams, or otherwise, describe a procedure to charge the sphere positively by induction. **2**

Marks

5. (continued)

(*c*) A charged oil drop of mass $1{\cdot}2 \times 10^{-14}$ kg is stationary between two horizontal parallel plates.

There is a potential difference of 4·9 kV between the parallel plates.

The plates are 80 mm apart as shown in Figure 7.

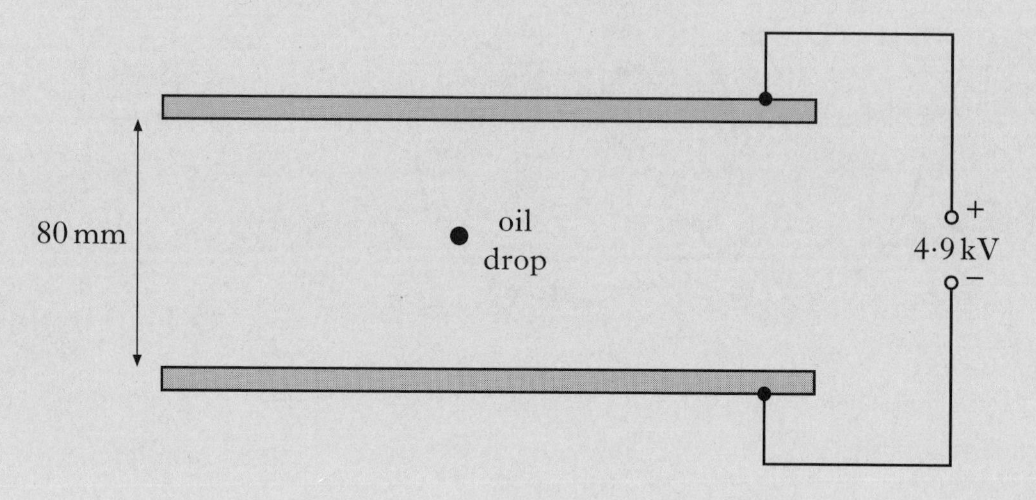

Figure 7

(i) Draw a labelled diagram to show the forces acting on the oil drop. **1**

(ii) Calculate the charge on the oil drop. **3**

(iii) How many excess electrons are on the oil drop? **1**

(*d*) The results of Millikan's oil drop experiment led to the idea of quantisation of charge.

A down quark has a charge of $-5{\cdot}3 \times 10^{-20}$ C. Explain how this may conflict with Millikan's conclusion. **1**

(11)

[Turn over

Marks

6. The shape of the Earth's magnetic field is shown in Figure 8.

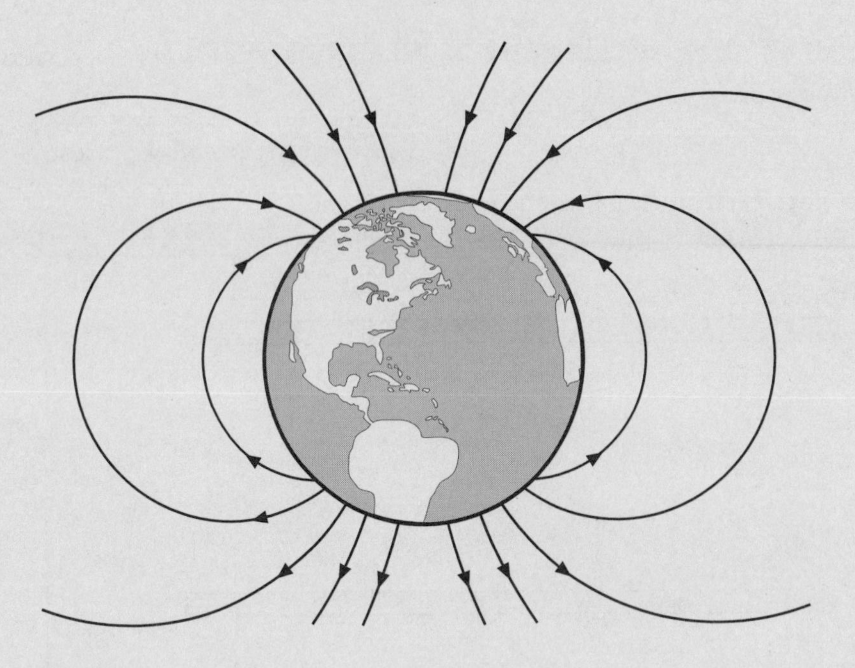

Figure 8

At a particular location in Scotland the field has a magnitude of $5{\cdot}0 \times 10^{-5}$ T directed into the Earth's surface at an angle of 69° as shown in Figure 9.

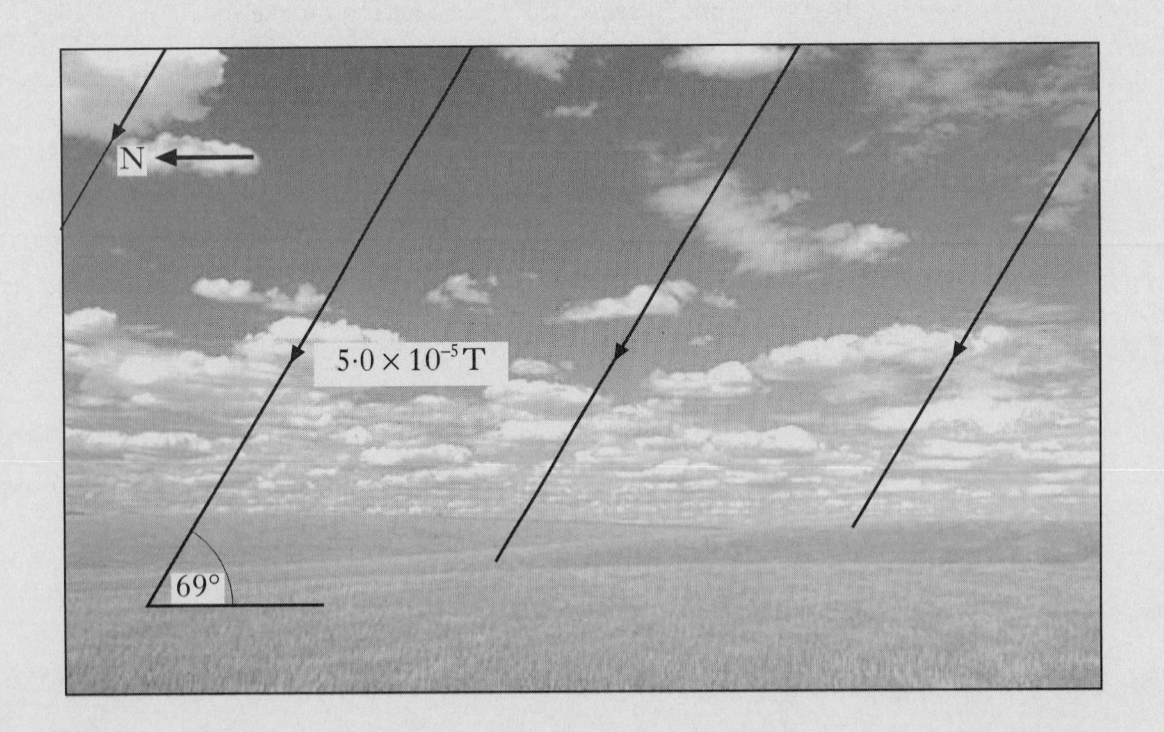

Figure 9

(*a*) Show that the component of the field perpendicular to the Earth's surface is $4{\cdot}7 \times 10^{-5}$ T. **1**

Marks

6. (continued)

(*b*) At this location a student sets up a circuit containing a straight length of copper wire lying horizontally in the North – South direction as shown in Figure 10.

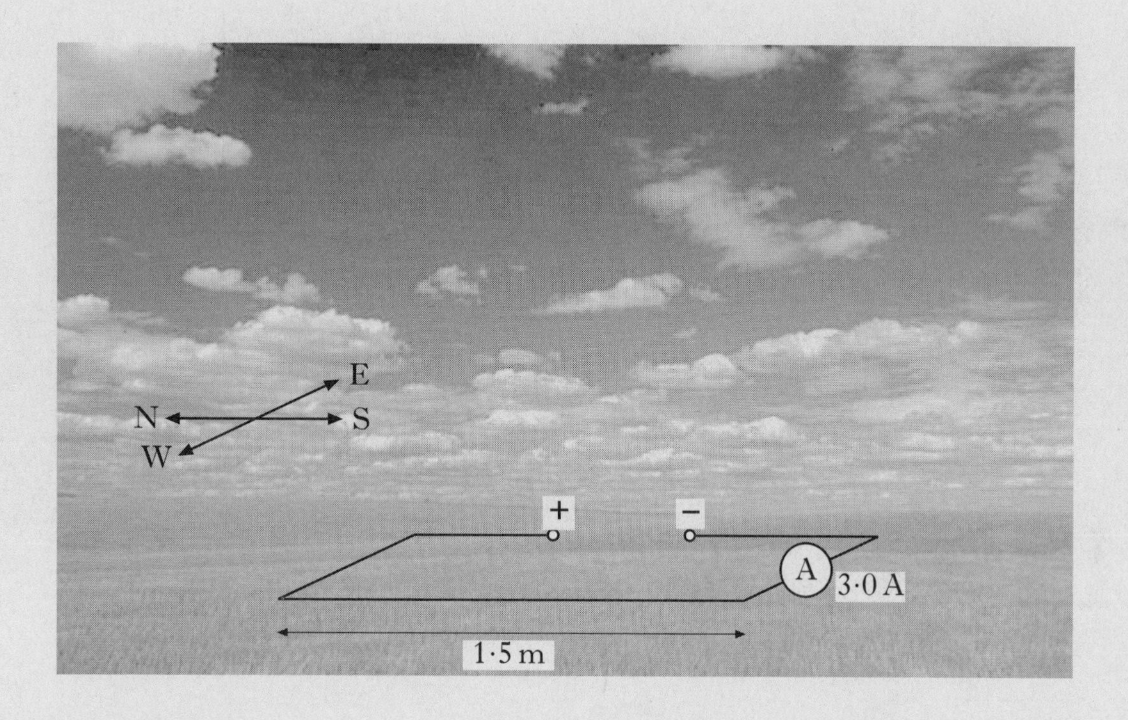

Figure 10

The length of the wire is 1·5 m and the current in the circuit is 3·0 A.

(i) Calculate the magnitude of the force acting on the wire due to the Earth's magnetic field. **2**

(ii) State the direction of this force. **1**

(*c*) The wire is now tilted through an angle of 69 ° so that it is parallel to the direction of the Earth's magnetic field.

Determine the force on the wire due to the Earth's magnetic field. **1**

(*d*) A long straight current carrying wire produces a magnetic field. The current in this wire is 3·0 A.

(i) Calculate the distance from the wire at which the magnitude of the magnetic field is $5{\cdot}0 \times 10^{-5}$ T. **2**

(ii) Describe the shape of this magnetic field. **1**

(8)

[Turn over

Marks

7. (*a*) Figure 11 shows a d.c. power supply in series with a switch, lamp and inductor.

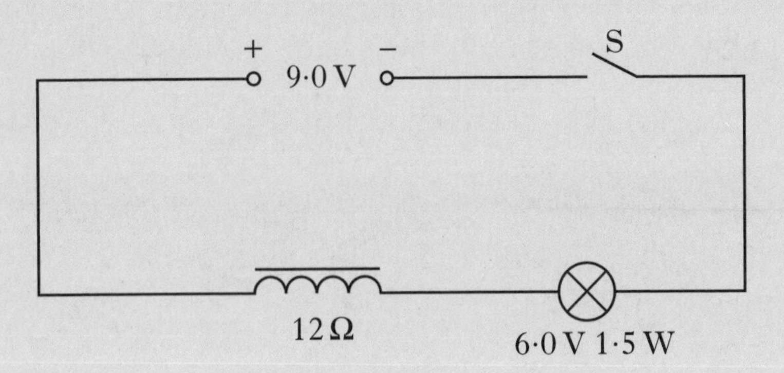

Figure 11

The inductor consists of a coil of wire with a resistance of 12 Ω.
The lamp is rated at 6·0 V 1·5 W.
The 9·0 V d.c. power supply has negligible internal resistance.

(i) Explain why the lamp does not reach its maximum brightness immediately after the switch is closed. **2**

(ii) When the lamp reaches its maximum brightness it is operating at its stated power rating.

Calculate the current in the circuit. **1**

(iii) The maximum energy stored in the inductor is 75 mJ.

Calculate the inductance of the inductor. **2**

(iv) The inductor in Figure 11 is replaced with another inductor which has the same type of core and wire, but with twice as many turns.

State the effect this has on:

(A) the maximum current;

(B) the time to reach maximum current. **2**

Marks

7. (continued)

(*b*) Figure 12 shows a neon lamp connected to an inductor, switch and a 1·5 V cell.

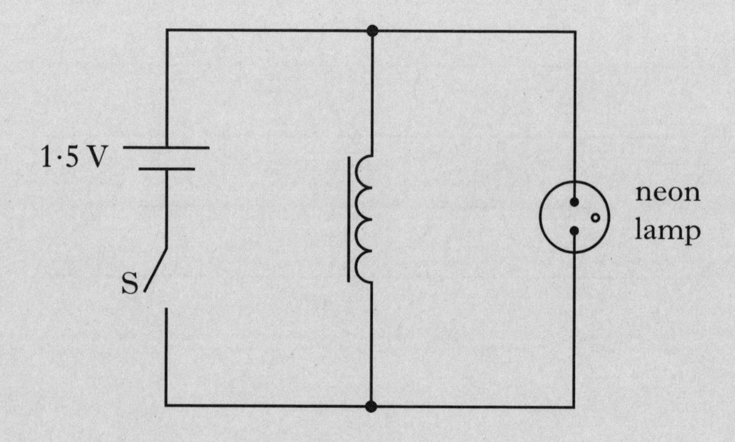

Figure 12

A neon lamp needs a potential difference of at least 80 V across it before it lights.
The switch is closed for 5 seconds.
The switch is then opened and the neon lamp flashes **briefly**.

Explain this observation. **2**

(9)

[Turn over

Marks

8. An electron travelling at $9{\cdot}5 \times 10^7\,\mathrm{m\,s^{-1}}$ enters a uniform magnetic field B at an angle of 60 ° as shown in Figure 13.

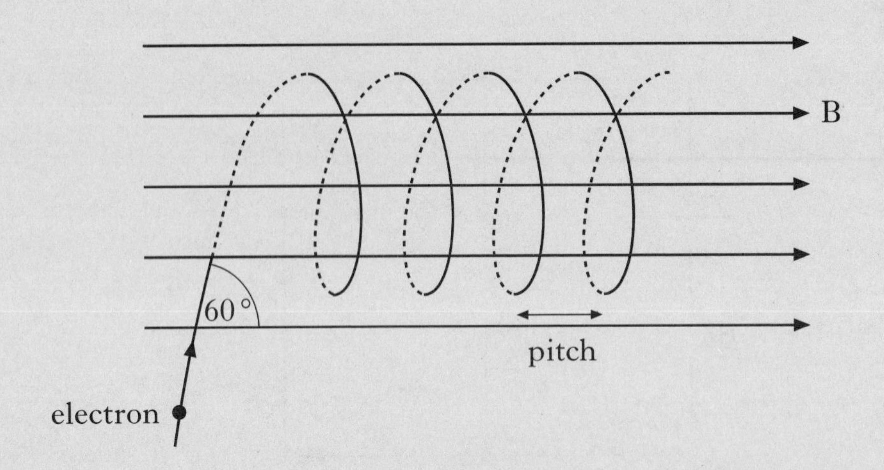

Figure 13

The electron moves in a helical path in the magnetic field.

(*a*) (i) Calculate the component of the electron's initial velocity:

(A) parallel to the magnetic field; **1**

(B) perpendicular to the magnetic field. **1**

(ii) By making reference to **both** components, explain why the electron moves in a helical path. **2**

(*b*) (i) The magnetic field has a magnetic induction of 0·22 T.

Show that the radius of the helix is $2{\cdot}1 \times 10^{-3}$ m. **2**

(ii) Calculate the time taken for the electron to make one complete revolution. **2**

(iii) The distance between adjacent loops in the helix is called the pitch as shown in Figure 13.

Calculate the pitch of the helix. **2**

(*c*) A proton enters the magnetic field with the same initial speed and direction as the electron shown in Figure 13. The magnetic field remains unchanged.

State **two** ways that the path of the proton in the magnetic field is different from the path of the electron. **2**

(12)

Marks

9. (*a*) A water wave travels with a speed of $0{\cdot}060\,\mathrm{m\,s^{-1}}$ in the positive x direction. Figure 14 represents the water wave at one instant in time.

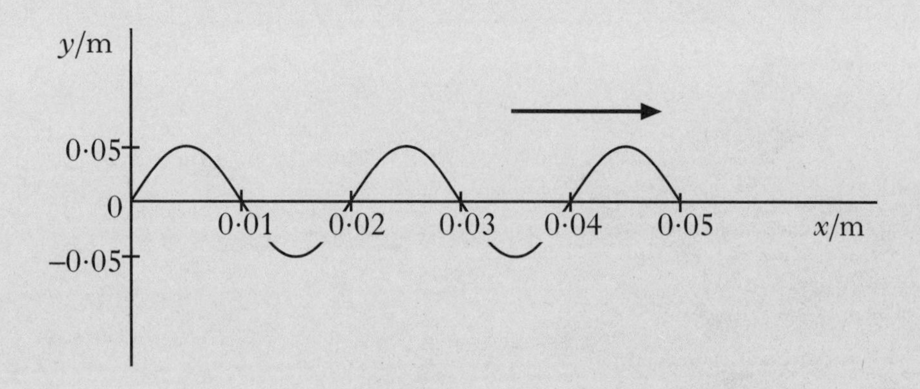

Figure 14

Write down an equation for the vertical displacement y of a point on the water surface in terms of the horizontal displacement x and time t.

Numerical values are required. **2**

(*b*) Write down an equation for an identical wave travelling in the opposite direction. **1**

(*c*) The amplitude of the wave gradually decreases.

Calculate the amplitude of the water wave when the intensity of the wave has decreased by 50%. **2**

(5)

[Turn over

Marks

10. (*a*) A thin coating of magnesium fluoride is applied to the surface of a camera lens.

Figure 15 shows an expanded view of this coating on the glass lens.

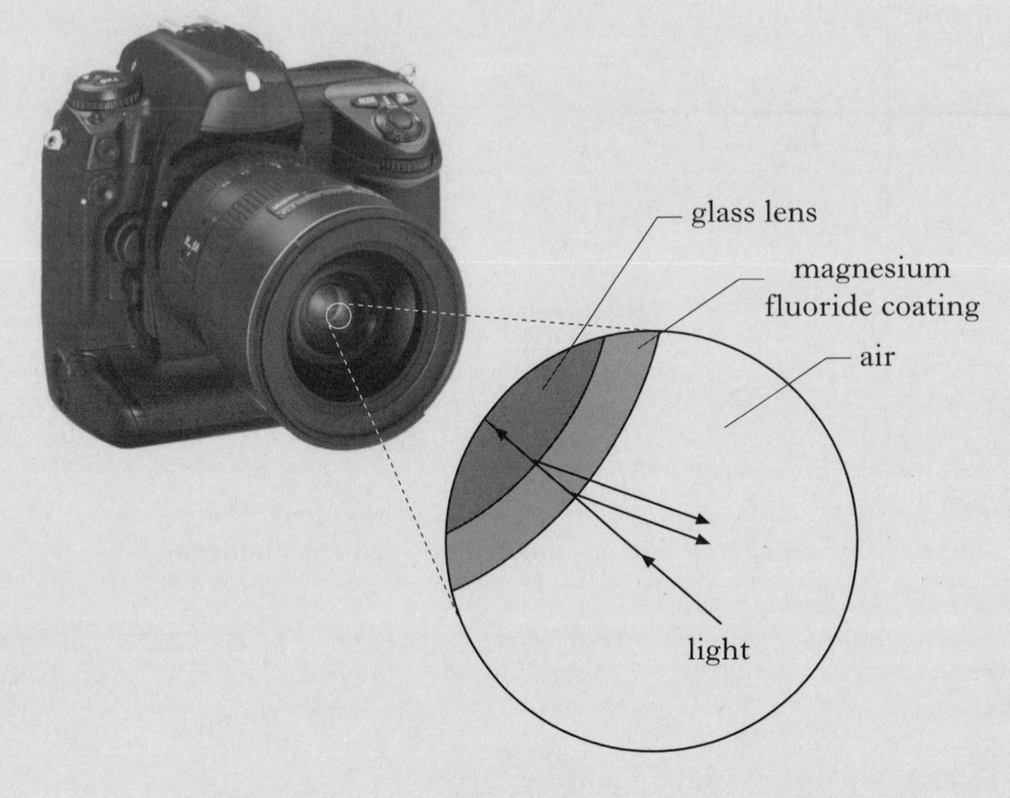

Figure 15

Monochromatic light is incident on the lens and some light reflects from the front and rear surfaces of the coating as shown in Figure 15.

(i) State the phase change undergone by the light reflected from:

(A) the front surface of the coating;

(B) the rear surface of the coating. **1**

(ii) Explain, in terms of optical path difference, why this coating can make the lens non-reflecting for a particular wavelength of light. **2**

(iii) Why is it desirable that camera lenses should reflect very little light? **1**

(iv) A particular lens has a magnesium fluoride coating of thickness $1{\cdot}05 \times 10^{-7}$m.

Calculate the wavelength of light for which this lens is non-reflecting. **2**

Marks

10. (continued)

(*b*) A thin air wedge is formed between two glass plates which are in contact at one end and separated by a thin metal wire at the other end.

Figure 16 shows sodium light being reflected down onto the air wedge. A travelling microscope is used to view the resulting interference pattern.

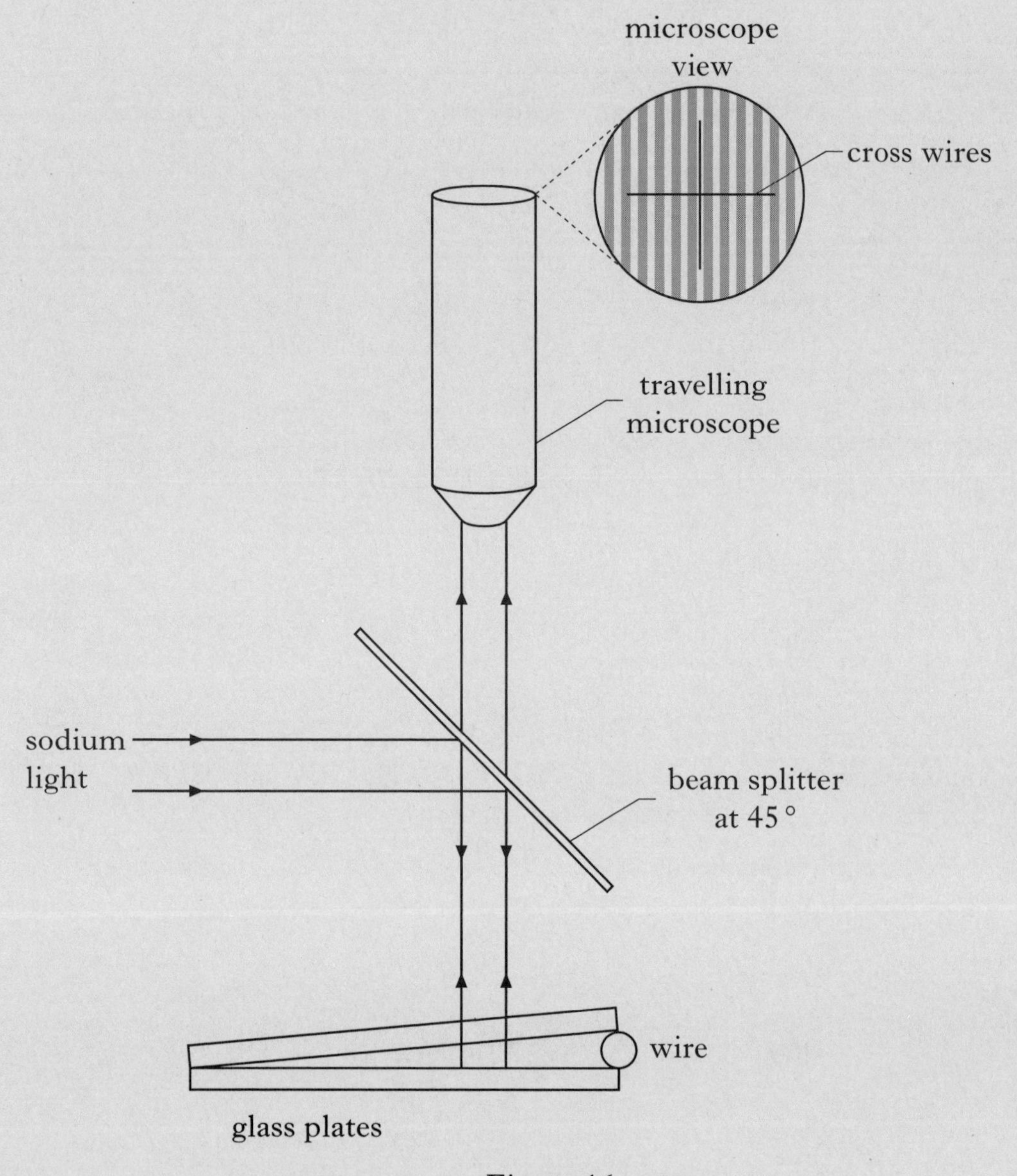

Figure 16

Explain how the diameter of the wire is determined using measurements obtained with this apparatus.

Assume the sodium light is monochromatic.

Your answer should include:

- the measurements required
- any data required
- the equation used.

2

(8)

Marks

11. The apparatus shown in Figure 17 is set up to measure the speed of transverse waves on a stretched string.

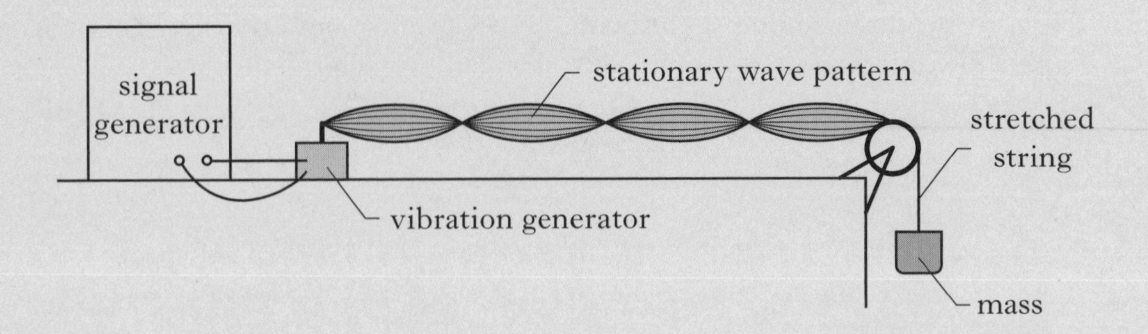

Figure 17

The following data are obtained.

Distance between adjacent nodes = (0·150 ± 0·005) m
Frequency of signal generator = (250 ± 10) Hz

(*a*) Show that the wave speed is 75 m s^{-1}. **2**

(*b*) Calculate the absolute uncertainty in this value for the wave speed. Express your answer in the form (75 ±   ) m s^{-1}. **3**

(*c*) (i) In an attempt to reduce the absolute uncertainty, the frequency of the signal generator is increased to (500 ± 10) Hz. Explain why this will **not** result in a reduced absolute uncertainty. **1**

(ii) State how the absolute uncertainty in wave speed could be reduced. **1**

(7)

[*END OF QUESTION PAPER*]

[BLANK PAGE]

X069/701

NATIONAL QUALIFICATIONS 2008

FRIDAY, 23 MAY
1.00 PM – 3.30 PM

PHYSICS
ADVANCED HIGHER

Reference may be made to the Physics Data Booklet.

Answer **all** questions.

Any necessary data may be found in the Data Sheet on page three.

Care should be taken to give an appropriate number of significant figures in the final answers to calculations.

Square-ruled paper (if used) should be placed inside the front cover of the answer book for return to the Scottish Qualifications Authority.

LI X069/701 6/4570

[BLANK PAGE]

DATA SHEET

COMMON PHYSICAL QUANTITIES

Quantity	*Symbol*	*Value*	*Quantity*	*Symbol*	*Value*
Gravitational acceleration on Earth	g	$9{\cdot}8\ \text{m s}^{-2}$	Mass of electron	m_e	$9{\cdot}11 \times 10^{-31}\ \text{kg}$
Radius of Earth	R_E	$6{\cdot}4 \times 10^{6}\ \text{m}$	Charge on electron	e	$-1{\cdot}60 \times 10^{-19}\ \text{C}$
Mass of Earth	M_E	$6{\cdot}0 \times 10^{24}\ \text{kg}$	Mass of neutron	m_n	$1{\cdot}675 \times 10^{-27}\ \text{kg}$
Mass of Moon	M_M	$7{\cdot}3 \times 10^{22}\ \text{kg}$	Mass of proton	m_p	$1{\cdot}673 \times 10^{-27}\ \text{kg}$
Radius of Moon	R_M	$1{\cdot}7 \times 10^{6}\ \text{m}$	Mass of alpha particle	m_α	$6{\cdot}645 \times 10^{-27}\ \text{kg}$
Mean Radius of Moon Orbit		$3{\cdot}84 \times 10^{8}\ \text{m}$	Charge on alpha particle		$3{\cdot}20 \times 10^{-19}\ \text{C}$
Universal constant of gravitation	G	$6{\cdot}67 \times 10^{-11}\ \text{m}^3\ \text{kg}^{-1}\ \text{s}^{-2}$	Planck's constant	h	$6{\cdot}63 \times 10^{-34}\ \text{J s}$
Speed of light in vacuum	c	$3{\cdot}0 \times 10^{8}\ \text{m s}^{-1}$	Permittivity of free space	ε_0	$8{\cdot}85 \times 10^{-12}\ \text{F m}^{-1}$
Speed of sound in air	v	$3{\cdot}4 \times 10^{2}\ \text{m s}^{-1}$	Permeability of free space	μ_0	$4\pi \times 10^{-7}\ \text{H m}^{-1}$

REFRACTIVE INDICES

The refractive indices refer to sodium light of wavelength 589 nm and to substances at a temperature of 273 K.

Substance	*Refractive index*	*Substance*	*Refractive index*
Diamond	2·42	Glycerol	1·47
Glass	1·51	Water	1·33
Ice	1·31	Air	1·00
Perspex	1·49	Magnesium Fluoride	1·38

SPECTRAL LINES

Element	*Wavelength*/nm	*Colour*
Hydrogen	656	Red
	486	Blue-green
	434	Blue-violet
	410	Violet
	397	Ultraviolet
	389	Ultraviolet
Sodium	589	Yellow
Cadmium	644	Red
	509	Green
	480	Blue

Lasers

Element	*Wavelength*/nm	*Colour*
Carbon dioxide	9550 10590	Infrared
Helium-neon	633	Red

PROPERTIES OF SELECTED MATERIALS

Substance	*Density*/ kg m^{-3}	*Melting Point*/ K	*Boiling Point*/ K	*Specific Heat Capacity*/ J kg^{-1} K^{-1}	*Specific Latent Heat of Fusion*/ J kg^{-1}	*Specific Latent Heat of Vaporisation*/ J kg^{-1}
Aluminium	$2{\cdot}70 \times 10^{3}$	933	2623	$9{\cdot}02 \times 10^{2}$	$3{\cdot}95 \times 10^{5}$	....
Copper	$8{\cdot}96 \times 10^{3}$	1357	2853	$3{\cdot}86 \times 10^{2}$	$2{\cdot}05 \times 10^{5}$	....
Glass	$2{\cdot}60 \times 10^{3}$	1400	....	$6{\cdot}70 \times 10^{2}$	....	....
Ice	$9{\cdot}20 \times 10^{2}$	273	....	$2{\cdot}10 \times 10^{3}$	$3{\cdot}34 \times 10^{5}$	....
Glycerol	$1{\cdot}26 \times 10^{3}$	291	563	$2{\cdot}43 \times 10^{3}$	$1{\cdot}81 \times 10^{5}$	$8{\cdot}30 \times 10^{5}$
Methanol	$7{\cdot}91 \times 10^{2}$	175	338	$2{\cdot}52 \times 10^{3}$	$9{\cdot}9 \times 10^{4}$	$1{\cdot}12 \times 10^{6}$
Sea Water	$1{\cdot}02 \times 10^{3}$	264	377	$3{\cdot}93 \times 10^{3}$	....	....
Water	$1{\cdot}00 \times 10^{3}$	273	373	$4{\cdot}19 \times 10^{3}$	$3{\cdot}34 \times 10^{5}$	$2{\cdot}26 \times 10^{6}$
Air	1·29	....	....	....	....	....
Hydrogen	$9{\cdot}0 \times 10^{-2}$	14	20	$1{\cdot}43 \times 10^{4}$	....	$4{\cdot}50 \times 10^{5}$
Nitrogen	1·25	63	77	$1{\cdot}04 \times 10^{3}$	....	$2{\cdot}00 \times 10^{5}$
Oxygen	1·43	55	90	$9{\cdot}18 \times 10^{2}$	....	$2{\cdot}40 \times 10^{5}$

The gas densities refer to a temperature of 273 K and a pressure of $1{\cdot}01 \times 10^{5}$ Pa.

Marks

1. A centrifuge is used to separate out small particles suspended in a liquid. Figure 1 shows the rotating part of the centrifuge which includes two test tubes containing the liquid.

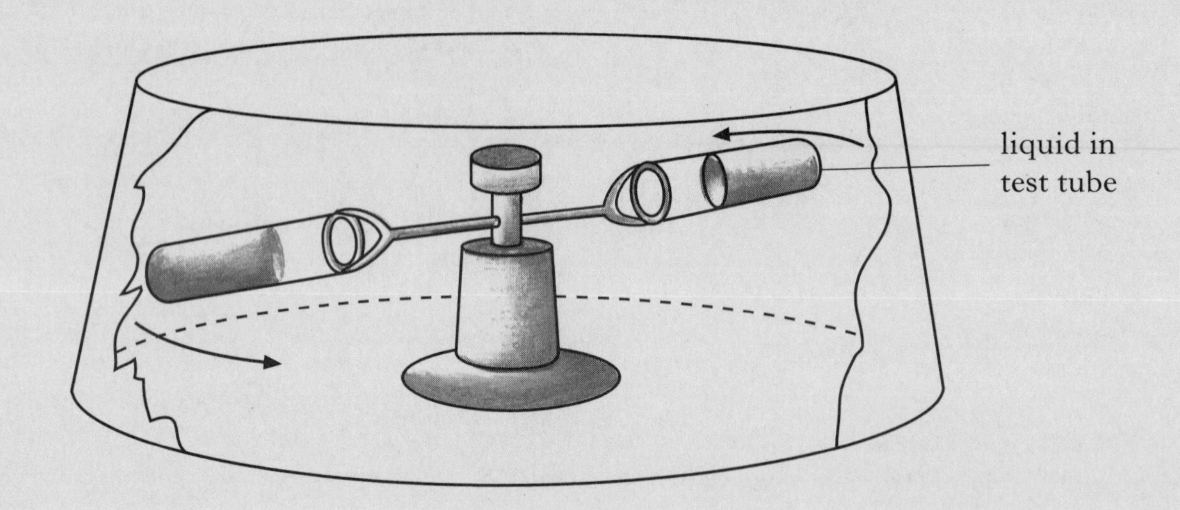

Figure 1

The rotating part starts from rest and reaches a maximum angular velocity of $1200\ \text{rad s}^{-1}$ in a time of 4 seconds.

The average moment of inertia of the rotating part is $5{\cdot}1 \times 10^{-4}\ \text{kg m}^2$.

(*a*) (i) Calculate the angular acceleration of the rotating part. **2**

(ii) Calculate the average unbalanced torque applied during this time. **2**

(iii) How many **revolutions** are made during this time? **3**

(*b*) Figure 2 shows an overhead view of the rotating part.

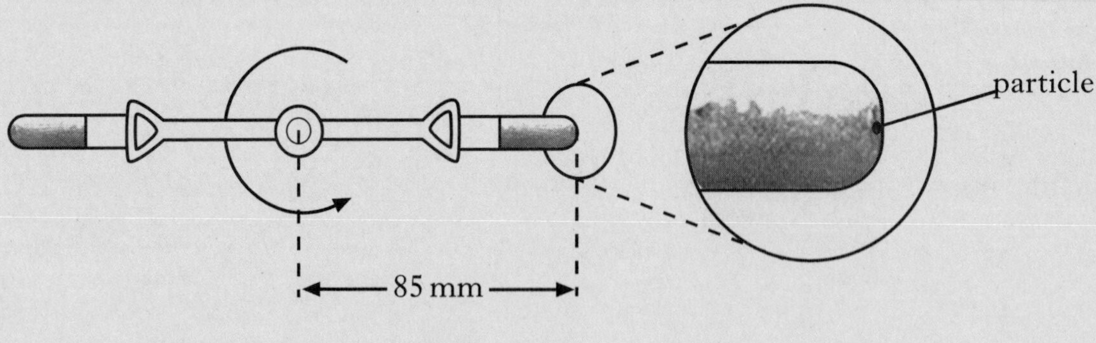

Figure 2

The expanded view shows the position of a single particle of mass $5{\cdot}3 \times 10^{-6}\ \text{kg}$.

(i) Calculate the central force acting on the particle at the maximum angular velocity. **2**

(ii) What provides the central force acting on this particle? **1**

Marks

1. (continued)

(*c*) At rest the test tubes in the centrifuge are in a vertical position as shown in Figure 3.

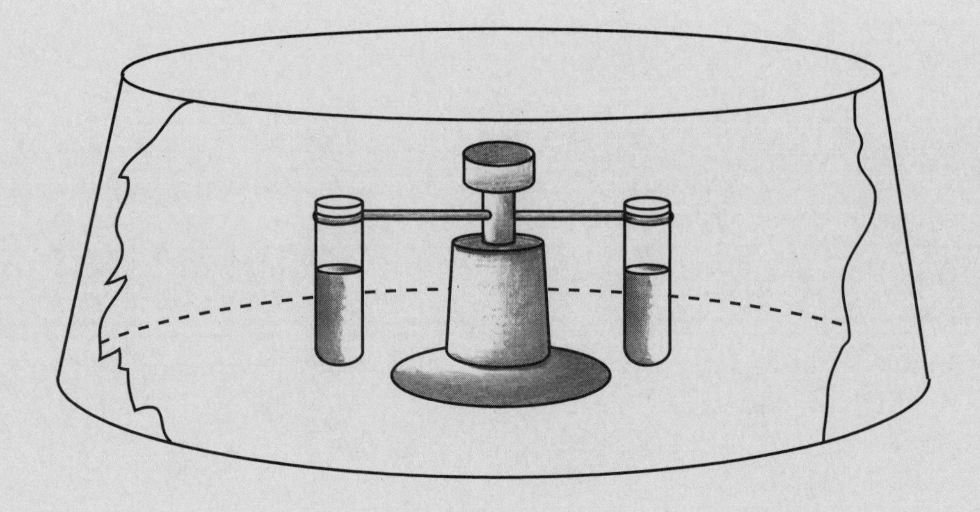

Figure 3

Does the moment of inertia of the rotating part increase, decrease, or stay the same during the acceleration of the rotating part? Justify your answer. **2**

(12)

[Turn over

Marks

2. (*a*) The gravitational field strength g on the surface of Mars is $3{\cdot}7\,\mathrm{N\,kg^{-1}}$.
The mass of Mars is $6{\cdot}4 \times 10^{23}\,\mathrm{kg}$.
Show that the radius of Mars is $3{\cdot}4 \times 10^{6}\,\mathrm{m}$. **2**

(*b*) (i) A satellite of mass m has an orbit of radius R. Show that the angular velocity ω of the satellite is given by the expression

$$\omega = \sqrt{\frac{GM}{R^3}}$$

where the symbols have their usual meanings. **2**

(ii) A satellite remains above the same point on the equator of Mars as the planet spins on its axis.

Figure 4 shows this satellite orbiting at a height of $1{\cdot}7 \times 10^{7}\,\mathrm{m}$ above the Martian surface.

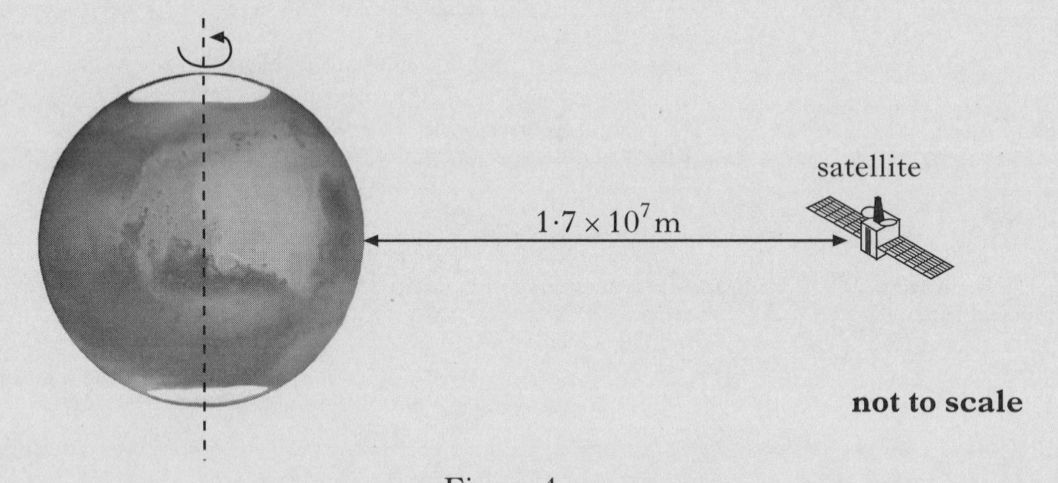

Figure 4

Calculate the angular velocity of the satellite. **2**

(iii) Calculate the length of one Martian day. **2**

(*c*) The following table gives data about three planets orbiting the Sun.

Planet	Radius R of orbit around the Sun/10^9 m	Orbit period T around the Sun/years
Venus	108	0·62
Mars	227	1·88
Jupiter	780	12·0

Use **all** the data to show that T^2 is directly proportional to R^3 for these three planets. **3**

(11)

Marks

3. A simple pendulum consists of a lead ball on the end of a long string as shown in Figure 5.

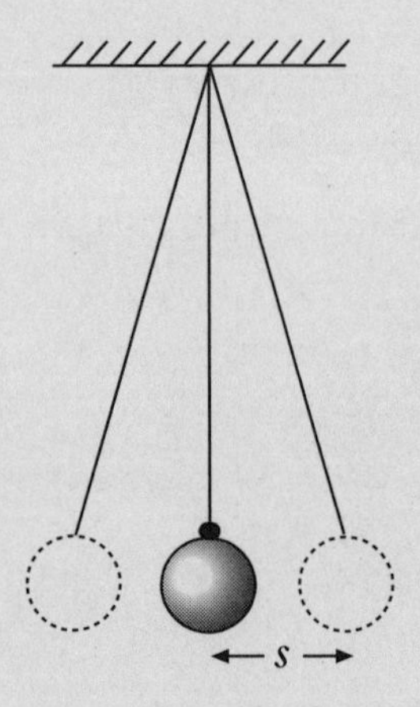

Figure 5

The ball moves with simple harmonic motion. At time t the displacement s of the ball is given by the expression

$$s = 2 \cdot 0 \times 10^{-2} \cos 4 \cdot 3t$$

where s is in metres and t in seconds.

(*a*) (i) State the definition of *simple harmonic motion*. **1**

(ii) Calculate the period of the pendulum. **2**

(*b*) Calculate the maximum speed of the ball. **2**

(*c*) The mass of the ball is $5 \cdot 0 \times 10^{-2}$ kg and the string has negligible mass.

Calculate the total energy of the pendulum. **2**

(*d*) The period T of a pendulum is given by the expression

$$T = 2\pi\sqrt{\frac{L}{g}}$$

where L is the length of the pendulum.

Calculate the length of this pendulum. **2**

(*e*) In the above case, the assumption has been made that the motion is not subject to *damping*.

State what is meant by *damping*. **1**

(10)

[Turn over

Marks

4. (*a*) Electrons can exhibit wave-like behaviour. Give **one** example of evidence which supports this statement. **1**

(*b*) The Bohr model of the hydrogen atom suggests a nucleus with an electron occupying one of a series of stable orbits.

A nucleus and the first two stable orbits are shown in Figure 6.

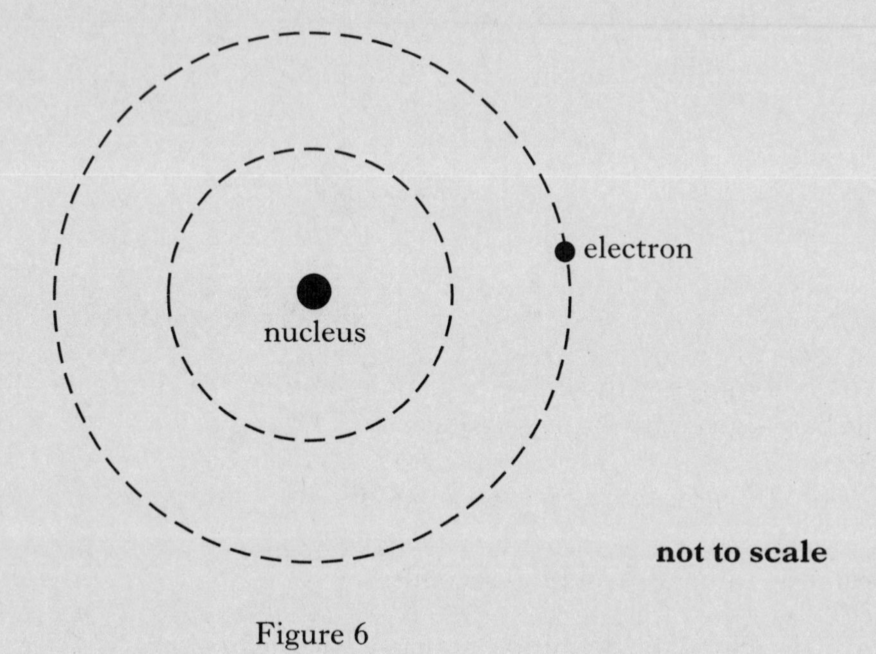

Figure 6

(i) Calculate the angular momentum of the electron in the second stable orbit. **2**

(ii) Starting with the relationship

$$mrv = \frac{nh}{2\pi}$$

show that the circumference of the second stable orbit is equal to two electron wavelengths. **2**

(iii) The circumference of the second stable orbit is $1{\cdot}3 \times 10^{-9}$ m.

Calculate the speed of the electron in this orbit. **2**

(7)

Marks

5. (*a*) Two point charges Q_1 and Q_2 each has a charge of $-4{\cdot}0\,\mu C$. The charges are $0{\cdot}60\,m$ apart as shown in Figure 7.

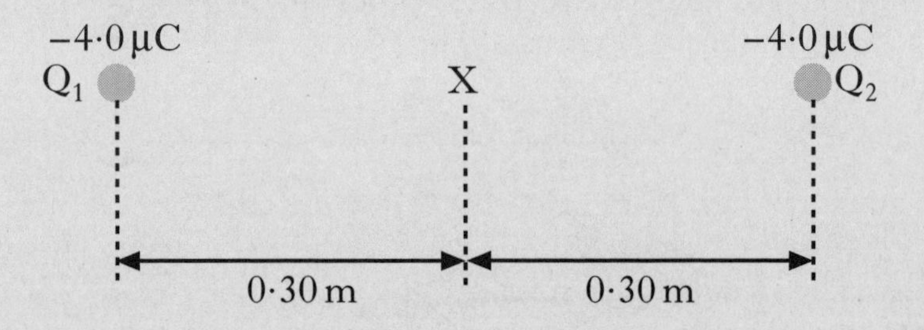

Figure 7

(i) Draw a diagram to show the electric field lines between charges Q_1 and Q_2. **1**

(ii) Calculate the electrostatic potential at point X, midway between the charges. **2**

(*b*) A third point charge Q_3 is placed near the two charges as shown in Figure 8.

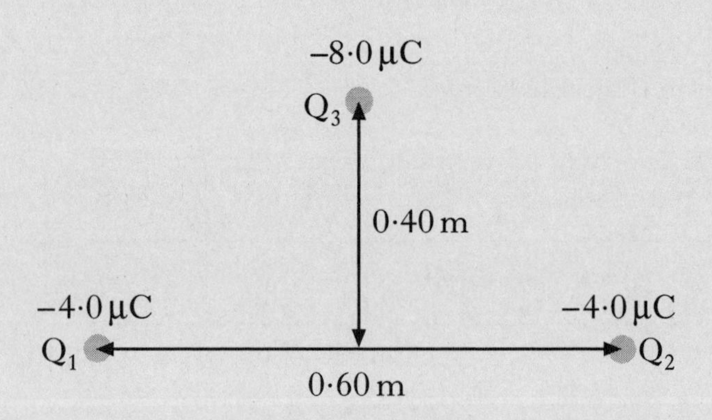

Figure 8

(i) Show that the force between charges Q_1 and Q_3 is $1{\cdot}2\,N$. **2**

(ii) Calculate the **magnitude** and **direction** of the resultant force on charge Q_3 due to charges Q_1 and Q_2. **2**

(7)

[Turn over

Marks

6. A student investigates the relationship between the force exerted on a wire in a magnetic field and the current in the wire.

A pair of magnets is fixed to a yoke and placed on a top pan Newton balance. A rigid copper wire is suspended between the poles of the magnets. The wire is fixed at 90° to the magnetic field, as shown in Figure 9.

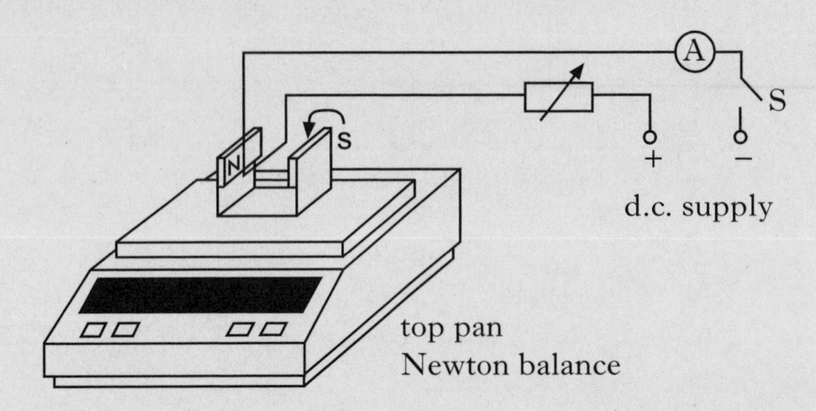

Figure 9

With switch S open the balance is set to zero.

Switch S is closed. The resistor is adjusted and the force recorded for several values of current.

The results are given in the table below.

Current/A	0·50	1·00	1·50	2·00	2·50
Force/10^{-3} N	0·64	0·85	2·56	3·07	3·87

The uncertainty in the current is ± 0·01 A.

The uncertainty in the force is $\pm\, 0{\cdot}03 \times 10^{-3}$ N.

Figure 10, on *Page eleven*, shows the corresponding graph with the best fit straight line for the results.

(*a*) (i) Show that the gradient of the line is $1{\cdot}7 \times 10^{-3}$ N A^{-1}. **1**

(ii) Calculate the absolute uncertainty in the gradient of the line. **3**

(iii) The length of wire in the magnetic field is 52 mm. Use the information obtained from the graph to calculate the magnitude of the magnetic induction.

The uncertainty in the magnetic induction is **not** required. **2**

(*b*) In the student's evaluation it is stated that the line does not pass through the origin.

(i) Suggest a possible reason for this. **1**

(ii) Suggest **one** improvement to the experiment to reduce the absolute uncertainty in the gradient of the line. **1**

(8)

6. (continued)

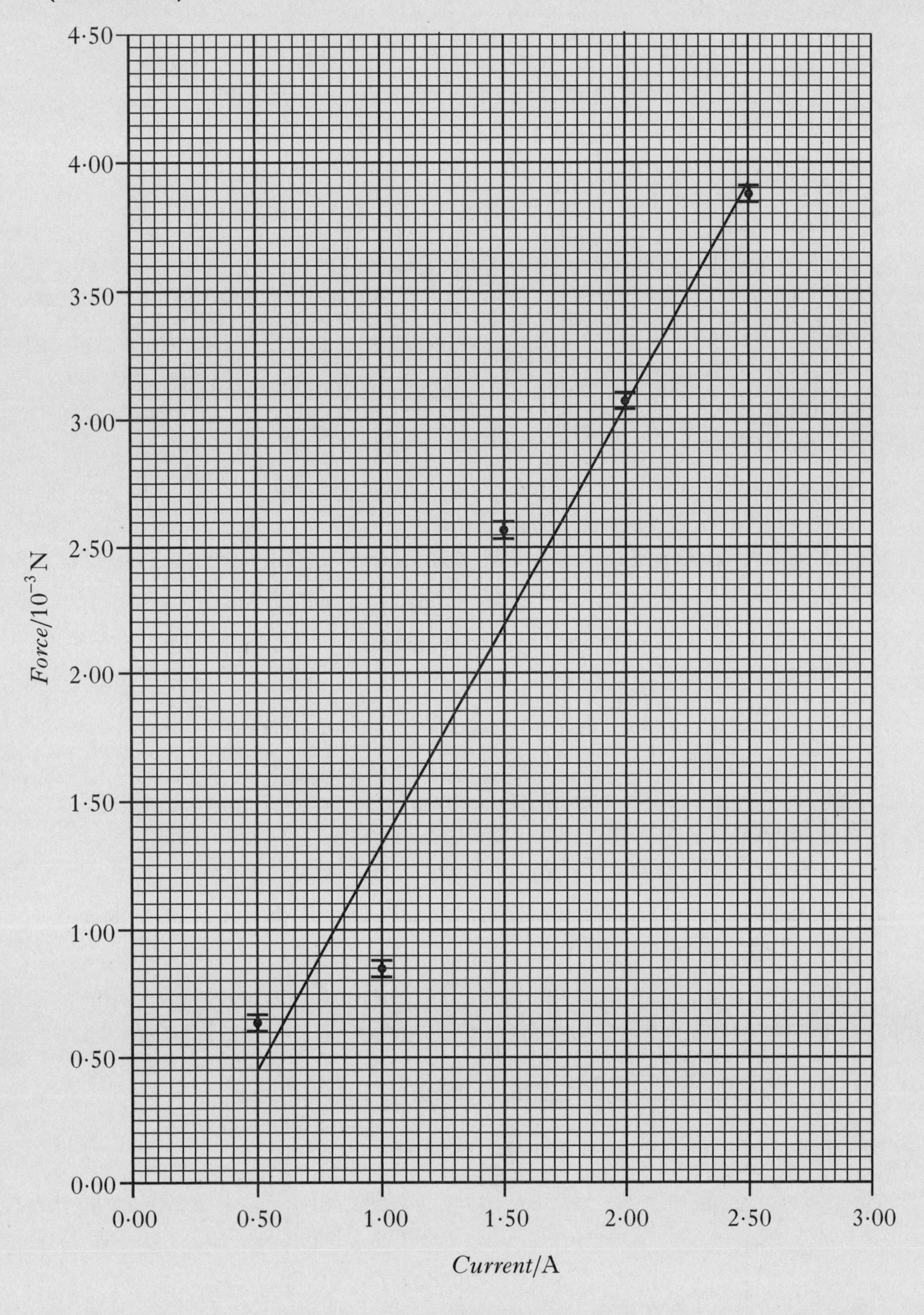

Figure 10

[Turn over

Marks

7. An inductor of negligible resistance is connected in the circuit shown in Figure 11.

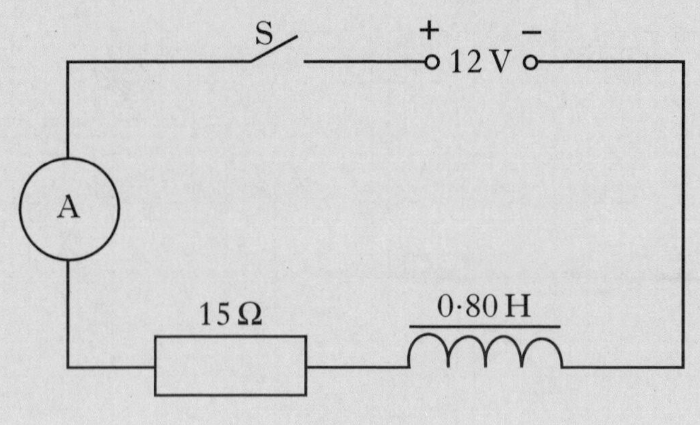

Figure 11

(*a*) The inductor has an inductance of 0·80 H.
Switch S is closed.

(i) Explain why there is a time delay before the current reaches its maximum value. **1**

(ii) Calculate the maximum current in the circuit. **2**

(iii) Calculate the maximum energy stored in the inductor. **2**

(iv) Calculate the rate of change of current when the current in the circuit is 0·12 A. **3**

(*b*) Switch S is opened and the iron core is removed from the inductor. Switch S is now closed.

(i) Will the maximum current be bigger, smaller or the same as the maximum current calculated in (*a*)(ii)? **1**

(ii) Explain any change in the time delay to reach the maximum current. **2**

(iii) Explain why the maximum energy stored in the inductor is less than in (*a*)(iii). **1**

(*c*) The iron core is replaced in the inductor. The d.c. supply is replaced with a variable frequency supply as shown in Figure 12.

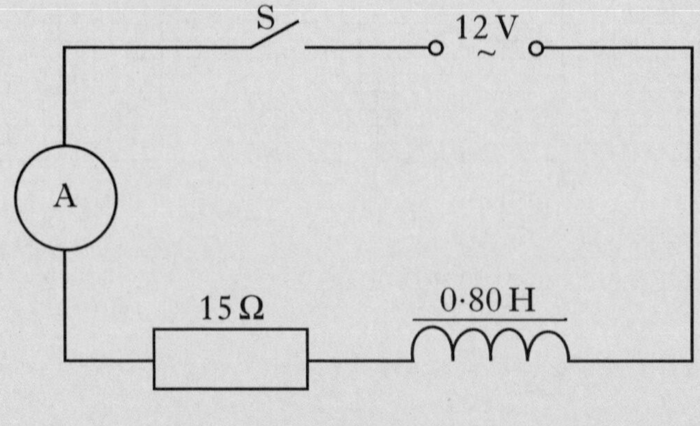

Figure 12

Sketch a graph to show how the current in the circuit varies with the frequency of the supply. Numerical values are not required. **1**

(13)

Marks

8. (*a*) Two protons are separated by a distance of 22 μm.

(i) Show by calculation that the gravitational force between these protons is negligible compared to the electrostatic force. **4**

(ii) Why is the strong force negligible between these protons? **1**

(*b*) A particle of charge q travels directly towards a fixed stationary particle of charge Q.

At a large distance from charge Q the moving particle has an initial velocity v.

The moving particle momentarily comes to rest at a distance of closest approach r_c as shown in Figure 13.

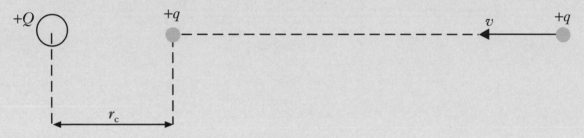

Figure 13

Show that the initial velocity of the moving particle is given by

$$v = \sqrt{\frac{qQ}{2\pi\varepsilon_0 m r_c}}$$

where the symbols have their usual meaning. **2**

(*c*) An alpha particle is fired towards a target nucleus which is fixed and stationary. The initial velocity of the alpha particle is $9{\cdot}63 \times 10^6\,\text{m s}^{-1}$ and the distance of closest approach is $1{\cdot}12 \times 10^{-13}\,\text{m}$.

(i) Calculate the charge on the target nucleus. **3**

(ii) Calculate the number of protons in the target nucleus. **2**

(iii) The target is the nucleus of an element. Identify this element. **1**

(13)

[Turn over

Marks

9. (*a*) The driver of a sports car approaches a building where an alarm is sounding as shown in Figure 14.

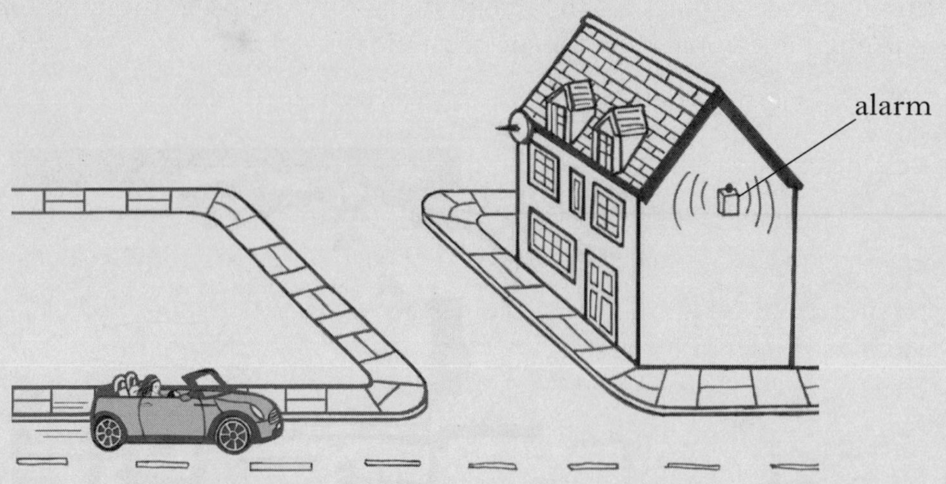

Figure 14

The speed of the car is $25{\cdot}0\,\mathrm{m\,s^{-1}}$ and the frequency of the sound emitted by the alarm is 1250 Hz.

(i) Explain in terms of wavefronts why the sound heard by the driver does not have a frequency of 1250 Hz. You may wish to include a diagram to support your answer. **2**

(ii) Calculate the frequency of the sound from the alarm heard by the driver. **2**

Marks

9. (continued)

(*b*) The spectrum of light from most stars contains lines corresponding to helium gas.

Figure 15(*a*) shows the helium spectrum from the Sun.

Figure 15(*b*) shows the helium spectrum from a distant star.

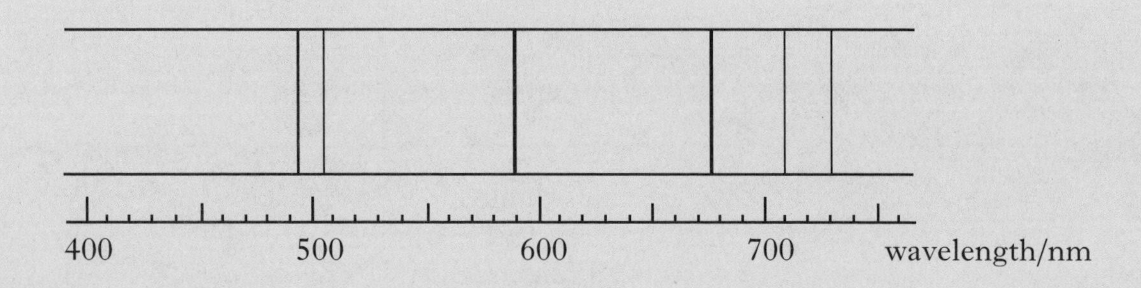

Figure 15(*a*)

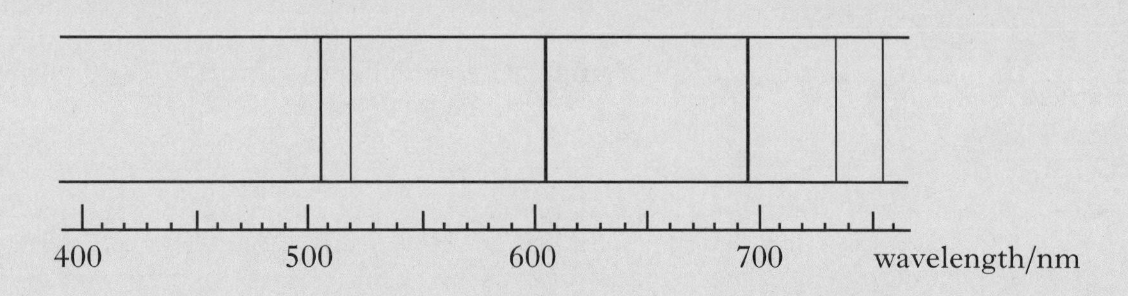

Figure 15(*b*)

By comparing these spectra, what conclusion can be made about the distant star? Justify your answer. **2**

(6)

[Turn over

Marks

10. (*a*) (i) State what is meant by the term *plane polarised light*. **1**

(ii) Figure 16 shows the refraction of red light at a water-air interface.

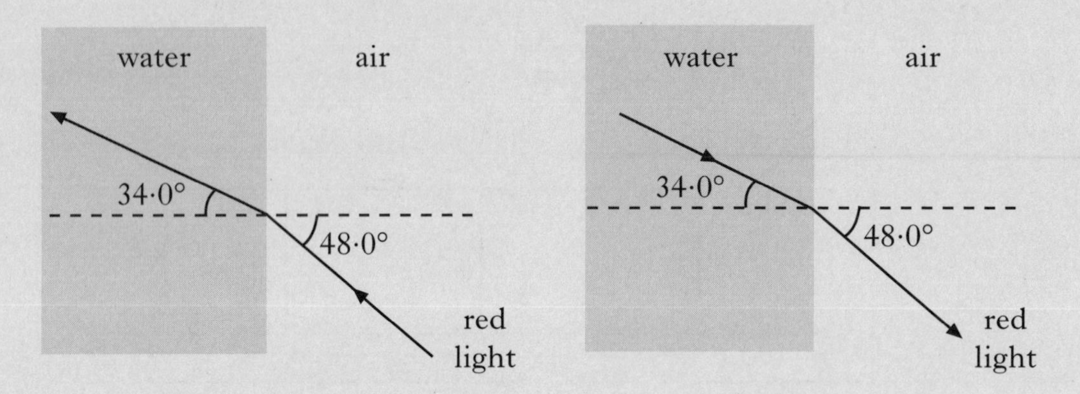

Figure 16

The refractive index n for red light travelling from air to water is 1·33. Show that the refractive index μ for red light travelling from **water** to **air** is 0·752. **1**

(iii) Figure 17 shows a ray of unpolarised red light incident on a water-air interface.

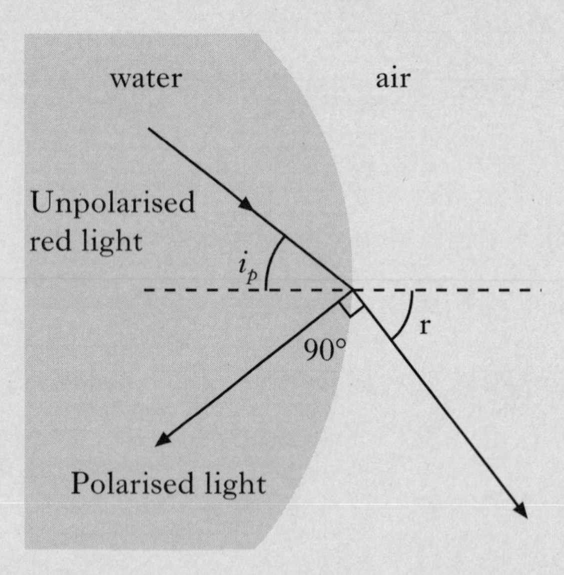

Figure 17

For light travelling from water to air,

$$\mu = \tan i_p$$

where i_p is the Brewster angle.

Calculate the Brewster angle for red light at this water-air interface. **1**

Marks

10. (continued)

(*b*) A rainbow is produced when light follows the path in a raindrop as shown in Figure 18.

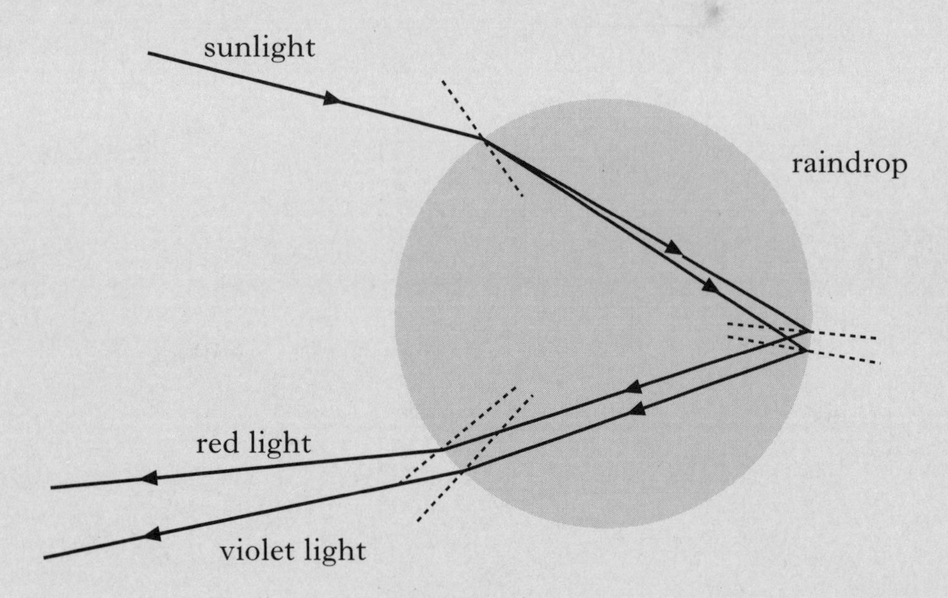

Figure 18

The light emerging from the raindrop is polarised.

The refractive index, μ, at a water to air interface is 0·752 for red light and 0·745 for violet light.

Calculate the difference in Brewster's angle for these two colours. **2**

(*c*) Rainbows produce light that is 96% polarised. A photographer plans to take a photograph of a rainbow. Her camera has a polarising filter in front of the lens as shown in Figure 19.

Figure 19

She directs her camera at the rainbow and slowly rotates the filter to see which is the best image to take.

Describe what happens to the image of the rainbow as she slowly rotates her filter through 180°. **2**

(7)

Marks

11. Light from a helium-neon laser is incident on a double slit. A pattern of light and dark fringes is observed on a screen 3·50 m beyond the slits as shown in Figure 20.

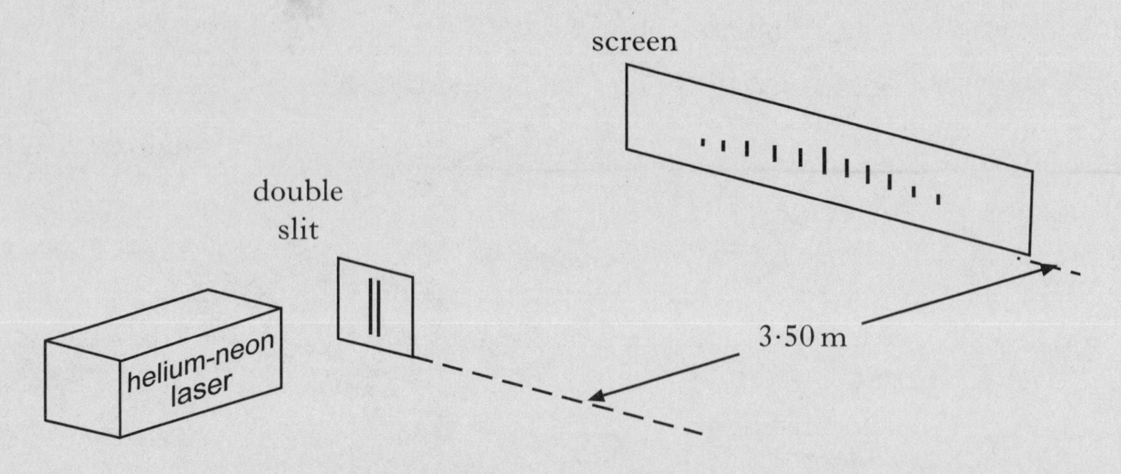

Figure 20

(*a*) State whether these fringes are caused by division of amplitude or division of wavefront. **1**

(*b*) The distance between two adjacent bright fringes on the screen is 7·20 mm. Calculate the separation of the two slits. **2**

(*c*) The distance between the double slit and screen is increased to 5·50 m. The distance between the fringes is remeasured and the calculation of the slit separation is repeated.

(i) Explain **one** advantage of moving the screen further away from the double slit. **2**

(ii) State **one** disadvantage of moving the screen further away from the double slit. **1**

(6)

[*END OF QUESTION PAPER*]

[BLANK PAGE]

[BLANK PAGE]

Pocket answer section for SQA Advanced Higher Physics 2004–2008

Published by Leckie & Leckie Ltd, 3rd Floor, 4 Queen Street, Edinburgh EH2 1JE
tel: 0131 220 6831, fax: 0131 225 9987, enquiries@leckieandleckie.co.uk, www.leckieandleckie.co.uk

Advanced Higher Physics 2004

1. (*a*) $2 \cdot 8 \times 10^8$ m/s

(*b*) $E = 1 \cdot 9 \times 10^{-13}$ J

2. (*a*) (i) (gravitational) Ep
rotational Ek
(linear/translational) Ek

(ii) rotational Ek

(*b*) $2 \cdot 5 \times 10^{-4}$ kg m^2

(*c*) $h = 0 \cdot 92$ m

(*d*) (i) The (5 N) force of spring is not great enough to provide the required centripetal force

(ii) $w = 204$ rad s^{-1}

3. (*a*) $w_0 = 20\pi$ (rad s^{-1})

(*b*) $\alpha = -2 \cdot 1$ rad s^{-2}
OR
deceleration $\alpha = 2 \cdot 1$ rad s^{-2}

(*c*) 150 (rev)

(*d*) 0·0045 Nm

(*e*) I less
α greater
time less than 30 s

4. (*a*)

$$\frac{GM_E m}{r^2} = mw^2 r$$
$$w = \frac{2\pi}{T}$$
$$\frac{GM_E m}{r^2} = \frac{m4\pi^2 r}{T^2}$$
$$T^2 = \frac{4\pi^2 r^3}{GM_E}$$
$$\left(T = 2\pi\sqrt{\frac{r^3}{GM_E}}\right)$$

(*b*) (i) $T = 2\pi\sqrt{\dfrac{r^3}{GM_E}}$

$$T = 2\pi\sqrt{\frac{(6 \cdot 4 \times 10^6 + 8 \times 10^4)^3}{6 \cdot 67 \times 10^{-11} \times 6 \times 10^{24}}}$$
$$T = 5180 \cdot 88 \text{ s}$$
$$\approx 86 \text{ min}$$

(ii) $2 \cdot 4 \times 10^6$ m

5. (*a*) $F = 2 \cdot 1$ N

(*b*) (i) "Restoring" force proportional to displacement OR $F = -kx$ OR $F \propto -x$

(ii) $a = 4 \cdot 2$ ms^{-2}

(iii) Amplitude = 0·06 m

5. (*c*) (i) Acceleration

(ii) $f = 1 \cdot 3$ Hz

6. (*a*) (i) The force per **unit positive** charge placed at a point in the field

(ii) $F = qE$
Work in moving charge between two plates
$= Fd = qEd$
$= qV$
so $V = Ed$ or $E = \dfrac{V}{d}$

(*b*) (i) & (ii)

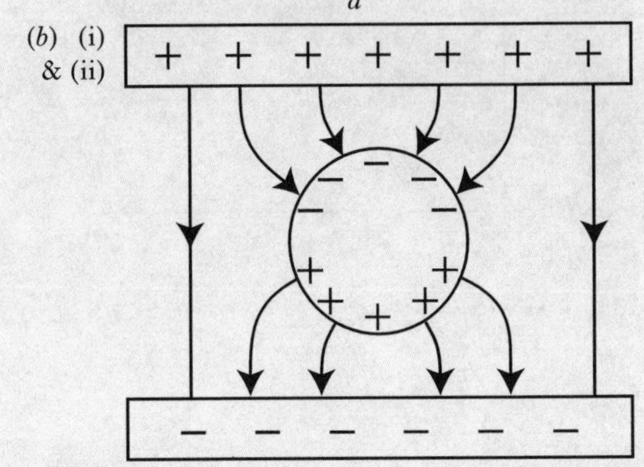

(iii) E = O

7. (*a*) $F = \dfrac{1}{4\pi\varepsilon_0} \times \dfrac{Q_1 Q_2}{r^2}$

(*b*) 1. Spheres touching
2. Bring positive rod near right of sphere S
3. Separate spheres

(*c*) (i)
$$F = \frac{1}{4\pi\varepsilon_0} \times \frac{Q_1 Q_2}{r^2}$$
$$3 \times 10^{-5} = \frac{q^2}{4\pi \times 8 \cdot 85 \times 10^{-12}(40 \times 10^{-3})^2}$$
$$q^2 = 5 \cdot 33819 \times 10^{-18}$$
$$(q = 2 \cdot 3 \times 10^{-9}\ C)$$

(ii) $V = 1700$ V

(iii)

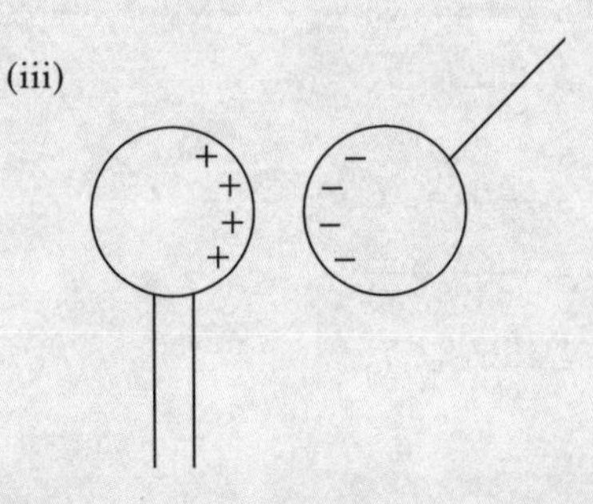

(*d*) (i) $T = 2 \cdot 5 \times 10^{-4}$ N

(ii) $\alpha = 7°$

Advanced Higher Physics 2004 (cont.)

8. (*a*) $\frac{mv^2}{r} = qvB$, so $r = \frac{mv}{qB}$

(*b*) $t = \frac{s}{v} \quad s = \pi r$

$$t = \frac{\pi r}{v} = \frac{\pi mv}{vqB} = \frac{\pi m}{qB}$$

(*c*) $3{\cdot}6 \times 10^{-6}$ ns

9. (*a*)

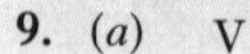

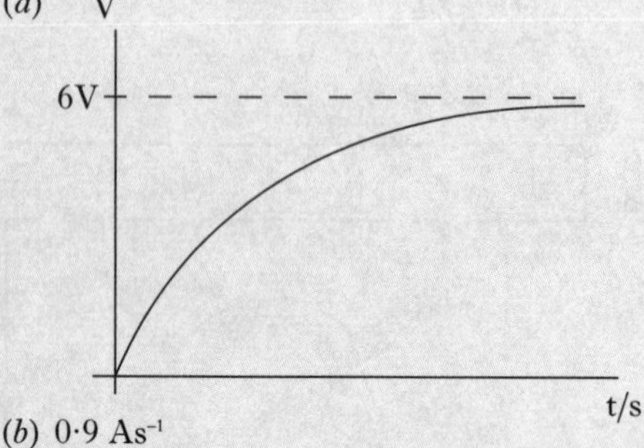

(*b*) $0{\cdot}9$ As^{-1}

10. (*a*) At wire 2

$$B = \frac{\mu_0 I_1}{2\pi r}$$

$$F = BI_2L$$

$$= \frac{\mu_0 I_1 I_2 L}{2\pi r}$$

$$\left(\frac{F}{L} = \frac{\mu_0 I_1 I_2}{2\pi r}\right)$$

(*b*) (i) $0{\cdot}036$ Nm^{-1}
(ii) apart
(iii) $B = 1{\cdot}7 \times 10^{-4}$ T upwards

(*c*) $0{\cdot}038$ Nm^{-1}

11. (*a*) (i) $1{\cdot}9$ Hz
(ii) $\lambda = 13$ m

(*b*) (i) $0{\cdot}48$ (rad)
(ii) $0{\cdot}04$°s

(*c*) $y = A \sin(12t + 0{\cdot}5x)$ where $A < 8$

12. (*a*) (i) The waves vibrate or oscillate in the same plane.
(ii) The picture will get poorer or disappear.

(*b*) (i) $n = \frac{\sin i_p}{\sin r}$

$$i_p + r = 90°$$

$$r = 90° - i_p$$

$$n = \frac{\sin i_p}{\sin (90 - i_p)}$$

$$n = \frac{\sin i_p}{\cos i_p} = \tan i_p$$

(ii) 56°

13. (*a*) $6{\cdot}76 \times 10^{-7}$ m

(*b*) $8{\cdot}6$%

(*c*) Increase D: this increases Δx and reduces uncertainty in Δx
OR
Measure more fringes: this reduces % uncertainty in Δx
OR
use a travelling microscope to measure Δx: this reduces scale reading uncertainty.

(*d*) Division of wavefront

Note: It is not essential to give the bracketed parts of these answers in your answers to obtain full marks. This is usually because they are given in the questions.

Advanced Higher Physics 2005

1. (*a*) 56·5 rad s^{-1}

(*b*) $\omega = \frac{v}{r}$

$= \frac{1 \cdot 3}{58 \times 10^{-3}}$

$= (22 \cdot 4 \text{ rad } s^{-1})$

(*c*) ωr is constant
r increases
(ω decreases)

(*d*) (i) θ = no. of revolutions $\times 2\pi$
$= 2 \cdot 8 \times 10^4 \times 2 \times 3 \cdot 14$
$= (1 \cdot 76 \times 10^5$ radians)

(ii) $- 7 \cdot 64 \times 10^{-3}$ $rads^{-2}$

(iii) 4460s (74·3 minutes)

2. (*a*) $I_{child} = mr^2$
$= 25 \times 2^2$
$= 100$ (kg m^2)

$I_{total} = I_{roundabout} + I_{child}$
$= 500 + 100$
$(= 600$ kg $m^2)$

(*b*) The angular momentum before (an impact) equals the angular momentum (after the impact) provided there are no external torques.

(*c*) (i) 60 kg m s^{-1}
(ii) 120 kg m^2 s^{-1}

(*d*) 0·2 rad s^{-1}

(*e*) 60 J

(*f*) (-) 3·8Nm

3. (*a*) (i) $-mR\omega^2 = \frac{GMm}{R^2}$

then substitute $\omega = \frac{2\pi}{T}$

(ii) $2 \cdot 4 \times 10^6$s

(*b*) (i) $E_P = \frac{-GM}{R} \times m$

$= \frac{-6 \cdot 67 \times 10^{-11} \times 6 \times 10^{24} \times 900}{(6 \cdot 4 \times 10^6 + 400 \times 10^3)}$

$(= -5 \cdot 3 \times 10^{10}\text{J})$

(ii) $E_K = \frac{1}{2}mv^2 = 2 \cdot 67 \times 10^{10}$ J

$E_{total} = E_p + E_k$
$= -5 \cdot 3 \times 10^{10} + 2 \cdot 67 \times 10^{10}$
$= -2 \cdot 6 \times 10^{10}$ J

4. (*a*) Acceleration is proportional to displacement (from a fixed point) and is always directed to (that) fixed point.
OR
The unbalanced force is proportional to the displacement (from a fixed point) and is always directed to (that) fixed point.
OR
$F = -kx$ where x = displacement
k = constant
F = force

(*b*) (i) 99·5 Hz (100Hz)
(ii) 8×10^{-4}m

(*c*) acceleration $= 9 \cdot 8$ m s^{-2}
(contact lost) when cap accelerates downwards greater (or equal to) g or similar.

5. (*a*) (i) $1 \cdot 8 \times 10^{10}$NC^{-1}
(ii) to the right

(*b*) (i) (A)

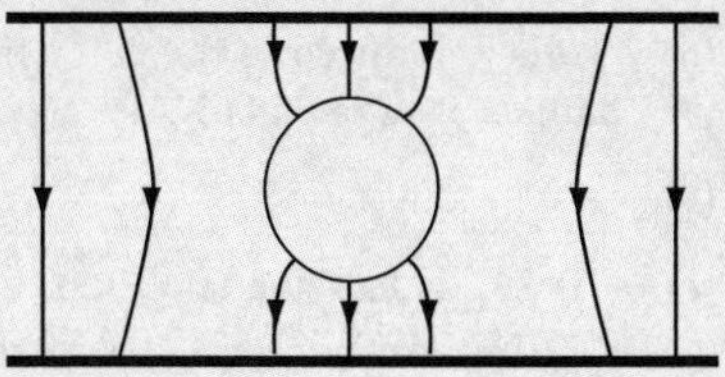

(B) 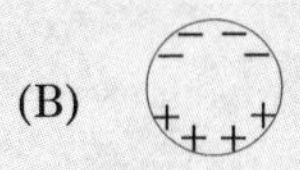

(ii) (external) electric fields (interference) cannot reach the central wire
(idea of shielding acceptable)

6. (*a*) $QV = \frac{1}{2}mv^2$
$1 \cdot 6 \times 10^{-19} \times 1500 = 0 \cdot 5 \times 9 \cdot 11 \times 10^{-31} \times v^2$
$v^2 = 5 \cdot 27 \times 10^{14}$
$(v = 2 \cdot 3 \times 10^7$ m $s^{-1})$

(*b*) $3 \cdot 9 \times 10^{-9}$s

(*c*) (i) $E = \frac{V}{d}$

$= \frac{600}{50 \times 10^{-3}}$

$= 12000$ (N C^{-1})
$F = EQ$
$= 12000 \times 1 \cdot 6 \times 10^{19}$
$(= 1 \cdot 9 \times 10^{-15}$N)

(ii) $1 \cdot 6 \times 10^{-2}$m

(*d*) (i) There is an unbalanced force (on the electron) in the vertical direction
(ii) No (unbalanced) forces act (on the electron)

(*e*) s increases
since v decreases
t decreases

Advanced Higher Physics 2005 (cont.)

7. (*a*) (i) positive
(ii) $9{\cdot}6 \times 10^{-7}$ C kg^{-1}
(iii) proton

$$\frac{q}{m} = \frac{1{\cdot}6\times10^{-19}}{1{\cdot}67\times10^{-27}}$$
$= 9{\cdot}6 \times 10^{-7}$ (C kg^{-1})

(*b*) The component of the electron's velocity perpendicular to the magnetic field causes circular motion (or equivalent).
The component of the electron's velocity parallel to the magnetic field is unchanged (or no force on electron parallel to B).

(*c*) Particles enter toward the poles.
Move in circles/spirals
OR
never reach atmosphere above equator

8. (*a*) 2 volts is induced in the coil when the current changes at (a rate of) 1 As^{-1}.

(*b*) 6 As^{-1}

(*c*) (i) I_{max} is less due to V_s less
(initial) $\frac{dI}{dt}$ is greater since L is smaller
(ii) 4·7 J

(*d*) (i) (gold bracelet) moves in magnetic field (conductor)
(ii) moving magnetic field or changing current induces current (voltage) in coil 2.

9. (*a*) (i) 250 Hz
(ii) $y = 4 \times 10^{-4}\sin(1570t + 4{\cdot}6x)$

(*b*) (i) Frequency greater than 800 Hz approaching, less than 800 HZ after train passes.
(ii) The waves (wavefronts) are closer together as they approach the person then they are further apart after they pass the person.

(iii) 18m s^{-1}

10. (*a*) (i) $5{\cdot}1 \times 10^{-7}$ m

(ii) % uncertainty in x = $\frac{0{\cdot}5\times100}{8}$ = 6·25%

% uncertainty in d = $\frac{0{\cdot}01\times100}{0{\cdot}25}$ = 4%

(% uncertainty in D = $\frac{0{\cdot}01\times100}{3{\cdot}91}$ = 0·3%)

% uncertainty in $\lambda = \sqrt{6{\cdot}25^2+4^2}$ = 7·4%
absolute certainty in $\lambda = 7{\cdot}4\% \times 5{\cdot}1 \times 10^{-7}$
$= 4 \times 10^{-8}$m

10. continued

(iii) An uncertainty should be quoted to one significant figure

(*b*) 2×10^{-8} m

(*c*) (i) The % uncertainty in x is very small compared to the % uncertainty in d
OR
reducing 0·8% still further does not change the uncertainty of λ at 4%.
(ii) the slit separation (d)

11. (*a*) Unpolarised light has the (electric field) oscillating in all planes.
Polarised light has the electric field oscillating in one plane only.

(*b*) (i) (Polarised) light cannot pass through the liquid crystal and is not reflected by the mirror.
(ii) Switch is opened.

(*c*) The numbers disappear (or cannot be seen) and then re-appear.
Light reflected from calculator is polarised
Indication of polarising material blocking/allowing transmission of light depending on rotation.

Advanced Higher Physics 2006

1. (*a*) (i) (A) $\overline{t} = \dfrac{\text{sum}}{\text{number}}$ OR

$$\overline{t} = \frac{10{\cdot}05 + 10{\cdot}30 + 9{\cdot}80 + 9{\cdot}97 + 9{\cdot}88}{5}$$

$$\overline{t} = \frac{50{\cdot}00}{5} = 10{\cdot}00$$

$$\overline{T} = \frac{10{\cdot}00}{20} = 0{\cdot}500\text{s}$$

(B) $\text{uncertainty} = \dfrac{\text{max - min}}{\text{N}}$

$$\text{uncertainty} = \frac{10{\cdot}30 - 9{\cdot}80}{5}$$

$$\text{uncertainty} = \frac{0{\cdot}500}{5} = 0{\cdot}100$$

$$\text{uncertainty in T} = \frac{0{\cdot}100}{20} = \pm 0{\cdot}005 \text{ s}$$

(ii) $F = mr\omega^2$

$$\omega = \frac{2\pi}{T} = \frac{2\pi}{0{\cdot}500} = 4\pi = (12{\cdot}56)$$

$$F = 0{\cdot}200 \times 0{\cdot}500 \times (4\pi)^2$$

$$F = 15{\cdot}8\text{N}$$

(iii) $\%\text{ uncertainty in r} = \dfrac{0{\cdot}002}{0{\cdot}500} \times 100 = 0{\cdot}4\%$

$$\%\text{ uncertainty in m} = \frac{0{\cdot}008}{0{\cdot}200} \times 100 = 4\%$$

$$\%\text{ uncertainty in T} = \frac{0{\cdot}005}{0{\cdot}500} \times 100 = 1\%$$

$$\%\text{ uncertainty in } T^2 = 2 \times 1 = 2\%$$

$$\%\text{ uncertainty in F} = \sqrt{(4^2 + 2^2)}$$

$$= 4{\cdot}47\%$$

$$F = (15{\cdot}8 \pm 0{\cdot}7)\text{N}$$

(iv) (A)

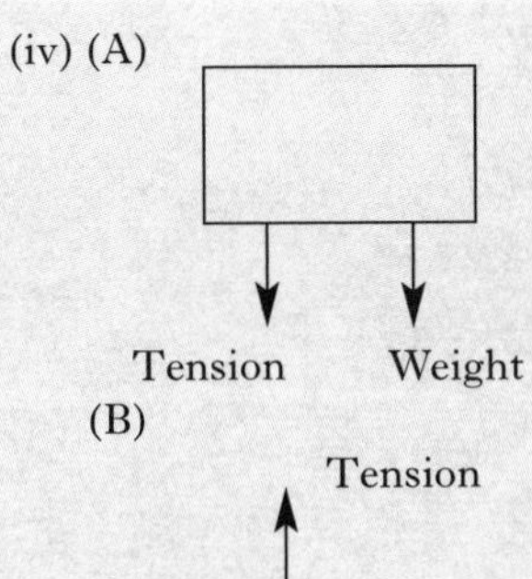

(B)

Tension

Weight

(v) weight = mg = $0{\cdot}200 \times 9{\cdot}8 = 1{\cdot}96$ (N)

$T_{min} = F_c - W = 15{\cdot}8 - 1{\cdot}96 = 13{\cdot}8$ N

(*b*) Pitch decreases
OR
Sound is lower

2. (*a*) (i) $\tau = F \times r$

$$\tau = 10 \times 20 \times 10^{-3}$$

$$\tau = 0{\cdot}20 \text{ Nm}$$

(ii) $\alpha = \dfrac{\omega - \omega_0}{t}$

$$\alpha = \frac{7{\cdot}5 - 0}{3}$$

$$\alpha = 2{\cdot}5 \text{ rads}^{-2}$$

(iii) $\tau = I \times \propto$

$$0{\cdot}20 = I \times 2{\cdot}5$$

$$I = \frac{0{\cdot}20}{2{\cdot}5} = 0{\cdot}080 \text{ kgm}^2$$

(*b*) (i) $I = {}^1/_2 mr^2$

$$I = 0{\cdot}5 \times 3{\cdot}2 \times (0{\cdot}12)^2$$

$$I = 0{\cdot}023 \text{ kgm}^2$$

(ii) $I_1\omega_1 = (I_1 + I_2)\omega_2$

$$0{\cdot}080 \times 7{\cdot}5 = (0{\cdot}080 + 0{\cdot}023)\omega_2$$

$$\omega_2 = 5{\cdot}8 \text{ rads}^{-1}$$

(*c*) Resulting angular velocity LESS
I_2 of the ring GREATER
Angular momentum CONSERVED

3. (*a*) (i) Force exerted on 1kg (of mass) placed in the field

(ii) (A) $F = mg$

$$mg = \frac{GmM}{r^2}$$

$$M = \frac{gr^2}{G}$$

(B) $M = \dfrac{3{\cdot}7 \times (3{\cdot}4 \times 10^6)^2}{6{\cdot}67 \times 10^{-11}}$

$$M = 6{\cdot}4 \times 10^{23} \text{ kg}$$

(*b*) (i) $F = \dfrac{GmM}{r^2}$

$$r = (3{\cdot}4 \quad 10^6 + 0{\cdot}3 \quad 10^6) = 3{\cdot}7 \quad 10^6$$

$$F = \frac{6{\cdot}67 \times 10^{-11} \times 100 \times 6{\cdot}4 \times 10^{23}}{(3{\cdot}7 \quad 10^6)^2}$$

$$F = 3{\cdot}1 \quad 10^2\text{N}$$

(ii) $F = mr\omega^2$

$$\omega^2 = \frac{3{\cdot}1 \times 10^2}{100 \times 3{\cdot}7 \times 10^6}$$

$$\omega = 9{\cdot}2 \times 10^{-4} (\text{rads}^{-1})$$

$$T = \frac{2\pi}{\omega} = \frac{2\pi}{9{\cdot}2 \times 10^{-4}}$$

$$T = 6{\cdot}8 \times 10^3\text{s}$$

Advanced Higher Physics 2006 (cont.)

4. (*a*) $E_k = {}^1/_2 m\omega^2(A^2 - y^2)$
$E_k = {}^1/_2 \times 0{\cdot}25 \times 36\ ((5\times 10^{-2})^2 - y^2)$
$E_k = 4{\cdot}5(2{\cdot}5 \times 10^{-3} - y^2)$

(*b*) $E_{k\,MAX} = 4{\cdot}5 \times 2{\cdot}5 \times 10^{-3}$
$E_{k\,MAX} = 1{\cdot}1 \times 10^{-2}J$

(*c*) $E_p = {}^1/_2 m\omega^2 y^2$
$E_p = {}^1/_2 \times 0{\cdot}25 \times 36 \times (0{\cdot}04)^2$
$E_p = 7{\cdot}2 \times 10^{-3}J$

5. (*a*) (i) $V = \frac{1}{4\pi\varepsilon_0} \cdot \frac{Q}{r}$ $r = 0{\cdot}18$ $\varepsilon_0 = 8{\cdot}85 \times 10^{-12}$

$$2{\cdot}8 \times 10^5 = \frac{Q}{4\pi \times 8{\cdot}85 \times 10^{-12} \times 0{\cdot}18}$$

$Q = 5{\cdot}6 \times 10^{-6}C$

(ii) $2{\cdot}8 \times 10^5 V$
(iii) (A)

$$E = \frac{1}{4\pi\varepsilon_0} \cdot \frac{Q}{r^2}$$
$$= \frac{5{\cdot}6 \times 10^{-6}C}{4\pi \times 8{\cdot}85 \times 10^{-12} \times 0{\cdot}18^2}$$
$$= 1{\cdot}6 \times 10^6 NC^{-1}$$

(B) E

r

(*b*) $r^2 = 0{\cdot}72$ $r = 0{\cdot}85$

$$E = \frac{1}{4\pi\varepsilon_0} \cdot \frac{Q}{r^2}$$
$$= \frac{5{\cdot}6 \times 10^{-6}C}{4\pi \times 8{\cdot}85 \times 10^{-12} \times 0{\cdot}85^2}$$
$$= 6{\cdot}99 \times 10^4$$

Vertical components sum to
$2 \times 6{\cdot}99 \times 10^4 \times \cos 45^o$
$= 9{\cdot}9 \times 10^4 NC^{-1}$
upwards

6. (*a*) $F = EQ$
$= 2{\cdot}5 \times 10^4 \times 1{\cdot}6 \times 10^{-12}$
$= 4{\cdot}0 \times 10^{-8}N$

(*b*) $W = mg$
$= 1{\cdot}2 \times 10^{-12} \times 9{\cdot}8$
$= 1{\cdot}2 \times 10^{-11}N$
<<Fe

6. (*c*) $a = \frac{F}{m}$
$$= \frac{4{\cdot}0 \times 10^{-8}}{1{\cdot}2 \times 10^{-12}}$$
$= (3{\cdot}3 \times 10^4 ms^{-2})$

$$t = \frac{l}{v} = \frac{7{\cdot}5 \times 10^{-3}}{20}$$
$= (3{\cdot}75 \times 10^{-4}s)$

$$s = (ut) + \frac{1}{2}at^2$$
$= 0{\cdot}5 \times 3{\cdot}3 \times 10^4 \times (3{\cdot}75 \times 10^{-4})^2$
$= 2{\cdot}3 \times 10^{-3}m$

(*d*) $n = \frac{Q}{e}$
$$= \frac{1{\cdot}6 \times 10^{-12}}{1{\cdot}60 \times 10^{-19}}$$
$= 1 \times 10^7$

7. (*a*) (i) $qV = {}^1/_2 mv^2$
$q = 1{\cdot}60 \times 10^{-19}$
$m = 1{\cdot}673 \times 10^{-27}$
$1{\cdot}60 \times 10^{-19} \times 2 \times 10^3 = 0{\cdot}5 \times 1{\cdot}673 \times 10^{-27} \times v^2$
$(v = 6{\cdot}19 \times 10^5\ ms^{-1})$

(ii) $Bqv = \frac{mv^2}{r} \Rightarrow r = \frac{mv}{Bq}$

$$= \frac{1{\cdot}673 \times 10^{-27} \times 6{\cdot}19 \times 10^5}{1{\cdot}3 \times 1{\cdot}60 \times 10^{-19}}$$
$$= 4{\cdot}98 \times 10^{-3}m$$

(iii) E_k doubled so v increased by $\sqrt{2}$
$v = 6{\cdot}19 \times 10^5 \times \sqrt{2}$
$= 8{\cdot}75 \times 10^5\ ms^{-1}$ or use
${}^1/_2\, mv^2 = {}^1/_2\, mu^2 + qV$

(*b*) $Bqv = \frac{mv^2}{r} \Rightarrow r = \frac{mv}{Bq}$

$$= \frac{1{\cdot}673 \times 10^{-27} \times 6{\cdot}19 \times 10^5}{1{\cdot}3 \times 1{\cdot}60 \times 10^{-19}}$$
$$= 1{\cdot}2 \times 10^{-13} m$$

(*c*) $m = \frac{m_0}{\sqrt{1 - \frac{v^2}{c^2}}}$

$$4{\cdot}66 \times 10^{-27} = \frac{1{\cdot}673 \times 10^{-27}}{\sqrt{1 - \frac{v^2}{(3 \times 10^8)^2}}}$$

$v = 2{\cdot}8 \times 10^8 ms^{-1}$

8. (*a*) Value of resistor R

(*b*) $E = -L \dfrac{dI}{dt}$

$-9 = -L \times 12$

$L = 0{\cdot}75$ H

(*c*) $E = {}^1/_2\, LI^2$

$= 0{\cdot}5 \times 0{\cdot}75 \times (0{\cdot}1)^2$

$= 3{\cdot}8 \times 10^{-3}$J

(*d*) $\dfrac{dI}{dt}$ **very** high

OR rapid decay or collapse of magnetic field

9. (*a*) (i) Strong (force)

(ii) A 2u + ld

B 1u + 2d

(*b*) weak (force)

(*c*) $\lambda = \dfrac{h}{mv}$

$= \dfrac{6{\cdot}63 \times 10^{-34}}{1{\cdot}675 \times 10^{-27} \times 3{\cdot}5 \times 10^{3}}$

$= 1{\cdot}13 \times 10^{-10}$m

10. (*a*) Nodes

(*b*) (i) $v = f\lambda$

$\lambda = 88 \times 10^{-3} \times 2$

$\lambda = 0{\cdot}176$(m)

$v = 2000 \times 0{\cdot}176$

$v = 350$ ms^{-1}

(ii) Measure distance between more than 2 nodes

or

Decrease frequency to increase λ

or

Decrease frequency to increase node separation

This will decrease uncertainty in measurement of λ

(*c*) Intensity of reflected sound wave now reduced

Difference between intensity of incident sound and reflected sound now greater

(Hence resultant intensity greater)

11. (*a*) (i)

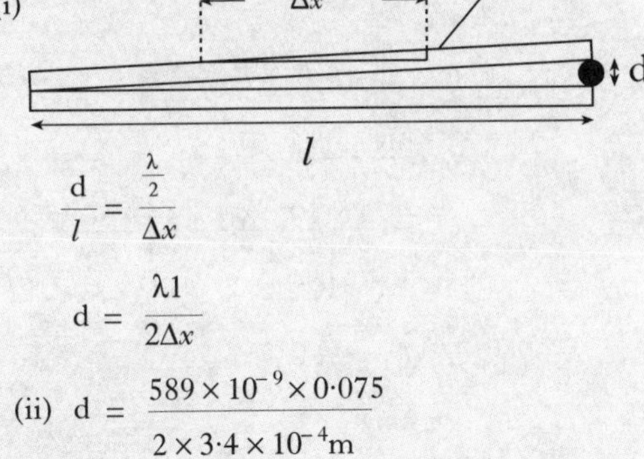

$\dfrac{d}{l} = \dfrac{\frac{\lambda}{2}}{\Delta x}$

$d = \dfrac{\lambda l}{2\Delta x}$

(ii) $d = \dfrac{589 \times 10^{-9} \times 0{\cdot}075}{2 \times 3{\cdot}4 \times 10^{-4}\text{m}}$

$= 6{\cdot}5 \times 10^{-5}$m

11. (*b*) (i) $d = \dfrac{\lambda}{4n}$

$= \dfrac{548 \times 10^{-9}}{4 \times 1{\cdot}38}$

$= 9{\cdot}9 \times 10^{-8}$m

(ii) $n_{MgF} < n_{Liquid}$

∴ no phase change at this surface

∴ constructive interference

∴ more light reflected

(*c*) (Path length) in oil depends on angle of incidence or thickness

different colours are seen due to interference

Advanced Higher Physics 2007

1. (*a*) $\frac{ds}{dt} = v$

$\int ds = \int (u + at) \cdot dt$
$s = ut + \frac{1}{2}at^2 + c.$
At $t = 0$, $s = 0$, so $c = 0$.
$\therefore s = ut + \frac{1}{2}at^2$

(*b*) $4{\cdot}2 \times 10^{-10}$ J

2. (*a*) (i) $6{\cdot}8 \times 10^{-3}$ kgm^2

(ii) A. $\omega = \frac{\text{no. of revs}}{60} \times 2\pi$
$= \frac{45}{60} \times 2\pi$
$= 4{\cdot}7$ rad s^{-1}

B. $3{\cdot}1$ rad s^{-2}

(iii) $3{\cdot}6$ rad s^{-1}

(iv) Centripetal force supplied by friction. Force of friction is less than the required centripetal force between mass and turntable.

(b) r (of arms and legs) decreases and so I decreases. Since $I\omega$ is constant, ω must increase.

3. (*a*) (i) $2{\cdot}0 \times 10^{20}$ N

(ii) $1{\cdot}0 \times 10^3$ ms^{-1}

(iii) work done in bringing unit mass from infinity (to a point in space)

(iv) $-7{\cdot}6 \times 10^{28}$ J

(v) $-4{\cdot}0 \times 10^{28}$ J

(*b*) (i) $E_P + E_K = 0$

$-\frac{GMm}{r} + \frac{1}{2}mv^2 = 0$

$\frac{1}{2}\cancel{m}v^2 = \frac{GM\cancel{m}}{r}$

$v = \sqrt{\frac{2GM}{r}}$

(ii) $2{\cdot}4 \times 10^3$ ms^{-1}

4. (*a*) Force (or acceleration) is proportional to displacement and directed towards centre.

(*b*) $y = 0{\cdot}050 \cos 628t$

(*c*) $2{\cdot}0 \times 10^4$ ms^{-2}

(*d*) $9{\cdot}6 \times 10^3$ N

(*e*) 240 J

5. (*a*) (i) $V = \frac{Q}{4\pi\varepsilon_0 r}$

$= \frac{5{\cdot}1 \times 10^{-9}}{4 \times 3{\cdot}14 \times 8{\cdot}85 \times 10^{-12} \times 0{\cdot}2}$

$= 230$ V

(ii) 140 V

(*b*) Bring rod close to sphere.
Earth (touch) sphere.
Remove earth before rod.

(*c*) (i)

electric force (upward arrow)

weight (or mg) (downward arrow)

(ii) $1{\cdot}92 \times 10^{-18}$ C

(iii) 12

(*d*) not an integer multiple of e

6. (*a*) $B_\perp = B \sin 69$
$= 5 \times 10^{-5} \times \sin 69$
$= 4{\cdot}7 \times 10^{-5}$ T

(*b*) (i) $2{\cdot}1 \times 10^{-4}$ N

(ii) east

(*c*) 0 N

(*d*) (i) $1{\cdot}2 \times 10^{-2}$ m

(ii) circular

7. (*a*) (i) Changing current causes changing magnetic field, producing back emf.

(ii) $0{\cdot}25$ A

(iii) $2{\cdot}4$ H

(iv) (A) smaller current
(B) longer time

(*b*) B collapses (rapidly) or $\frac{dI}{dt}$ is large.

Large (back) emf induced or emf > 80V

8. (*a*) (i) (A) $4{\cdot}8 \times 10^7$ m s^{-1}
(B) $8{\cdot}2 \times 10^7$ m s^{-1}

(ii) Perpendicular component results in circular motion/central force.
Parallel component results in constant velocity/ no horizontal force.

(*b*) (i) $\frac{mv^2}{r} = Bqv$

$\frac{m(v \sin\theta)^2}{r} = Bqv \sin\theta$

$r = \frac{mv \sin\theta}{Bq}$

$= \frac{9{\cdot}11 \times 10^{-31} \times 9{\cdot}5 \times 10^7 \times \sin 60}{0{\cdot}22 \times 1{\cdot}6 \times 10^{-19}}$

$= 2{\cdot}1 \times 10^{-3}$m

(ii) $1{\cdot}6 \times 10^{-10}$ s

(iii) $7{\cdot}6 \times 10^{-3}$ m

(*c*) *Any two from:*

- bigger radius
- spirals in opposite sense (direction)
- bigger pitch

9. (*a*) $y = 0{\cdot}05 \sin 2\pi\left(3t - \frac{x}{0{\cdot}02}\right)$

(*b*) $y = 0{\cdot}05 \sin 2\pi\left(3t + \frac{x}{0{\cdot}02}\right)$

(*c*) 0·04 m (0·035 m)

10. (*a*) (i) (A) π
(B) π
(ii) The reflected rays interfere destructively if optical path difference $= \frac{\lambda}{2}$
(iii) So more light is transmitted through the lens.
(iv) $5{\cdot}80 \times 10^{-7}$ m

(*b*) Measure:
- fringe separation
- length of glass plates

Required data:
- wavelength of sodium light

Equation used:
- $\Delta x = \frac{\lambda l}{2d}$

11. (*a*) $\lambda = 2 \times 0{\cdot}15$
$v = f\lambda$
$= 250 \times 0{\cdot}3$
$= 75 \text{ m s}^{-1}$

(*b*) $v = (75 \pm 4) \text{ m s}^{-1}$

(c) (i) % uncertainty in λ will increase.
(ii) Measure the distance over several nodes and take an average.

Advanced Higher Physics 2008

1. (*a*) (i) 300 rads^{-2}
(ii) 0·15 Nm
(iii) 380

(*b*) (i) 0·65 N
(ii) the glass tube

(*c*) I increases due to increase in r

2. (*a*) $(m)g = \frac{GM(m)}{R^2}$

$R^2 = \frac{6{\cdot}67 \times 10^{-11} \times 6{\cdot}4 \times 10^{23}}{3{\cdot}7}$

$= 1{\cdot}15 \times 10^{13}$

$R = \sqrt{1{\cdot}15 \times 10^{13}}$

$= 3{\cdot}4 \times 10^6$ m

(*b*) (i) $\frac{GMm}{R^2} = m\omega^2 R$

$\frac{GM}{R^2} = \omega^2 R$

$\omega^2 = \frac{GM}{R^2}$

(ii) $7{\cdot}1 \times 10^{-5} \text{ rads}^{-1}$
(iii) $8{\cdot}9 \times 10^4$ s

(*c*) $\frac{T^2}{R^3} = 3{\cdot}05 \times 10^{-7}$

$\frac{T^2}{R^3} = 3{\cdot}02 \times 10^{-7}$

$\frac{T^2}{R^3} = 3{\cdot}03 \times 10^{-7}$

statement $\frac{T^2}{R^3}$ = constant

or show graph of T^2 against R^3 is a straight line

3. (*a*) (i) acceleration α – displacement
or
force α displacement and directed towards a fixed point
(ii) 1·5 s

(*b*) $8{\cdot}6 \times 10^{-2} \text{ m s}^{-1}$

(*c*) $1{\cdot}8 \times 10^{-4}$ J

(*d*) 0·56 m

(*e*) Amplitude decreases

4. (*a*) Electrons exhibit { diffraction / interference

Advanced Higher Physics 2008 (cont.)

4. (*b*) (i) $= 2{\cdot}1 \times 10^{-34}$ kg m^2 s^{-1}

(ii) $mrv = \dfrac{nh}{2\pi}$

$mrv = \dfrac{2h}{2\pi}$

$2\pi r = \dfrac{2\,h}{mv}$

$= \dfrac{2h}{p}$

$= 2\lambda_B$

(iii) $= 1{\cdot}1 \times 10^6$ ms^{-1}

5. (*a*) (i)

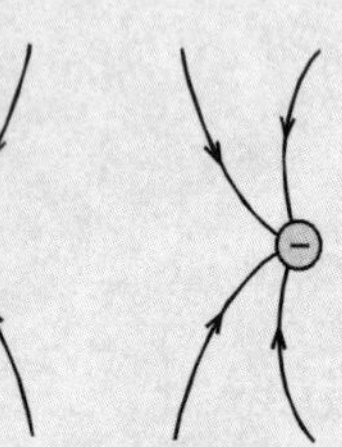

(ii) $-2{\cdot}4 \times 10^5$ V

(*b*) (i)

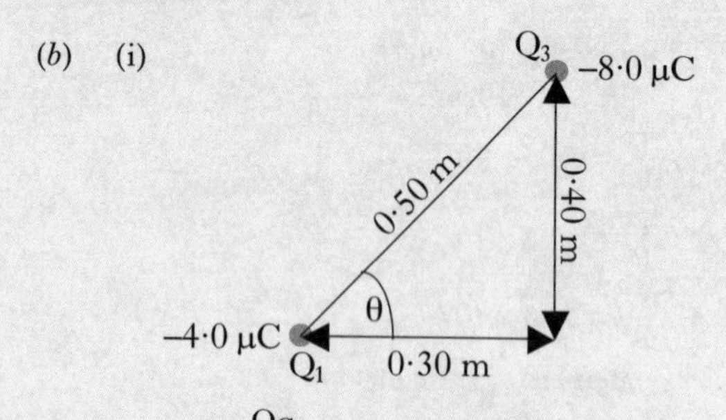

$F = \dfrac{Qq}{4\pi\varepsilon_0 r^2}$

$F = \dfrac{-8 \times 10^{-6} \times -4 \times 10^{-6}}{4 \times \pi \times 8{\cdot}85 \times 10^{-12} \times 0{\cdot}5^2}$

$= 1{\cdot}2$ N

(ii) F = 1·9 N
Direction (000°)

6. (*a*) (i) gradient $= \dfrac{y_2 - y_1}{x_2 - x_1}$

gradient $= \dfrac{3{\cdot}50 - 1{\cdot}50}{2{\cdot}25 - 1{\cdot}10}$

$= 1{\cdot}7 \times 10^{-3}$ NA^{-1}

(ii) $\pm 0{\cdot}3 \times 10^{-3}$ NA^{-1}
(Accept $(0{\cdot}2\text{–}0{\cdot}3) \times 10^{-3}$ NA^{-1}

(iii) $3{\cdot}3 \times 10^{-2}$T

(*b*) (i) Systematic uncertainty, calibration or zero error

(ii) *Any one from:*
- take more readings (to increase n)
- increase the range (to narrow parallelogram)
- take multiple readings and average

7. (*a*) (i) A changing/increasing current in the inductor generates a back emf.
(ii) 0·80 A
(iii) E = 0·26 J
(iv) 13 As^{-1}

(*b*) (i) unchanged
(ii) The time delay is decreased or the time to reach maximum current is reduced, because the inductance is decreased (by removing the iron core).
OR back emf is reduced
(iii) I is the same but L is smaller.

(*c*)

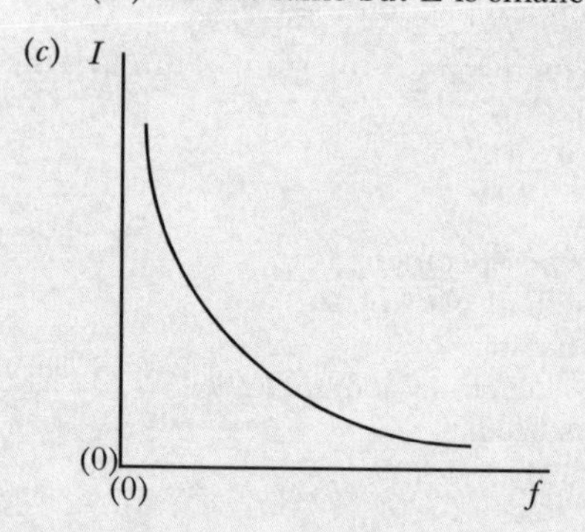

8. (*a*) (i) $F_{(E)}\ \dfrac{Q_1Q_2}{4\pi\varepsilon_0 r_2} = 4{\cdot}8 \times 10^{-19}$(N)

$F_{(G)} = \dfrac{Gm_1m_2}{r_2}$

$= 3{\cdot}9 \times 10^{-55}$ (N)

$F_G(10^{-55}N) << F_E(10^{-19}N)$

(ii) Strong force only acts at a range of approx. 10^{-14} m.
OR
The distance between these 2 protons is too large.

(*b*) $E_K = E_p$

$\therefore \frac{1}{2}mv^2 = \dfrac{Qp}{4\pi\varepsilon_0 r_c}$

$v^2 = \dfrac{2Qq}{4\pi\varepsilon_0 m r_c}$

$v = \sqrt{\dfrac{Qq}{2\pi\varepsilon_0 m r_c}}$

(*c*) (i) $1{\cdot}2 \times 10^{-17}$ C
(ii) 75
(iii) Rhenium

9. (*a*) (i) frequency increased (with moving observer)
Driver passing through more (than 1250) wavefronts in 1 second
(ii) 1340 Hz

(*b*) Distant star is moving away since wavelength increased
OR frequency decreased

10. *(a)* (i) Polarised light: (The electric field vector of) the wave **<u>oscillates</u>** or **<u>vibrates</u>** in one **<u>plane</u>**.

(ii) $\mu = \frac{\sin i}{\sin r}$

$\mu = \frac{\sin 34{\cdot}0}{\sin 48{\cdot}0} = 0{\cdot}752$

OR

$\mu = \frac{1}{n}$

$\mu = \frac{1}{1{\cdot}33} = 0{\cdot}752$

(iii) 36·9°

(b) 0·2°

(c) Intensity changes
In a cyclic fashion

11. *(a)* Division of wavefront

(b) $3{\cdot}1 \times 10^{-4}$ m

(c) (i) increased Δx
Smaller <u>%</u> uncertainty in d or Δx

(ii) Fainter fringes
OR broader fringes
OR not all fringes seen, screen not big enough

Official SQA answers to 978-1-84372-694-4
2004–2008